CHEMISTRY IN DAILY LIFE

This book is based on the Chemistry Value Added Course (VAC) for B.Sc. First Semester under NEP-2020, following the Choice Based Credit System (CBCS) to enhance students' skills and knowledge beyond the core syllabus.

CHEMISTRY IN DAILY LIFE

This book is based on the Chemistry Value Added Course (VAC) for B.Sc. First Semester under NEP-2020, following the Choice Based Credit System (CBCS) to enhance students' skills and knowledge beyond the core syllabus.

Dr. Toyaj Shukla

red'shine
PUBLICATION
INDIA

CHEMISTRY IN DAILY LIFE

by: Dr. Toyaj Shukla

■

RED'SHINE PUBLICATION PVT. LTD.

Headquarters (India): 88-90 REDMAC, Navamuvada,
Lunawada, India-389 230
Contact: +91 76988 26988
Registration no. GJ31D0000034

In Association with,

RED'MAC INTERNATIONAL PRESS & MEDIA. INC

India | Sweden | UK

■

■

■

ISBN: 978-93-5879-889-0
ISBN-10: 93-5879-889-0
DIP: 18.10.9358798890
DOI: 10.25215/9358798890
Price: ₹ 400
Editon: August, 2025 (First Edition)

■

■

www.redshine.co.in | info@redshine.in
Printed in India | Title ID: 9358798890

PREFACE

With profound gratitude, I begin by thanking the Almighty for His divine blessings that guided me in completing this textbook. I am deeply thankful to my parents, Family whose love, values, and constant encouragement have always inspired me. My heartfelt thanks to my beloved wife, whose patience, support, and motivation remained a strong pillar throughout this journey. I also extend sincere appreciation to the Principal Dr Sudheer Kumar Singh and all faculty members of Government Rani Durgawati College, whose valuable guidance, academic support, and encouragement played a vital role in shaping and completing this book.

Special thanks to my friends and colleagues, who offered moral support, reviewed my drafts, and helped me remain focused. Lastly, I acknowledge the silent blessings of well-wishers and mentors, without whom this academic endeavor would not have been successful. Ding and interesting path towards some sort of Edited Book without having their help and support, none of this work could have been possible.

Author

CONTENTS

CHAPTER 1

FOOD CHEMISTRY AND INDUSTRIAL APPLICATIONS

Learning Objectives

Upon completion of this chapter, students will be able to:

- Understand the fundamental chemical composition of milk and dairy products and apply analytical techniques for quality assessment
- Analyze the chemical constituents of beverages and identify adulterants using appropriate analytical methods
- Evaluate food preservatives, additives, and artificial sweeteners through chemical analysis
- Assess colorants, contaminants, and pesticide residues in food systems using modern instrumental techniques • Comprehend the chemical principles underlying paints, pigments, and industrial dyes

Introduction

Food chemistry represents one of the most fundamental applications of chemical principles in everyday life, encompassing the analysis, preservation, and quality control of consumable products. This field has evolved dramatically since the pioneering work of **Justus von Liebig** in the mid-19th century, who first applied systematic chemical analysis to food components. The discipline now integrates sophisticated analytical techniques with traditional methods to ensure food safety, authenticity, and nutritional value.

The chemical complexity of food systems presents unique challenges that require a deep understanding of both organic and inorganic chemistry. Modern food analysis must address concerns ranging from **nutritional adequacy** to **safety assessment**, including the detection of adulterants, contaminants, and residual substances. The integration of advanced instrumental techniques such as high-performance liquid chromatography (HPLC), gas chromatography-mass spectrometry (GC-MS), and atomic absorption spectroscopy (AAS) has revolutionized our ability to characterize food components at trace levels.

This chapter explores the intricate relationship between chemical composition and food quality, beginning with the foundational analysis of dairy products and extending through beverage chemistry, food additives, and industrial applications. The convergence of **food chemistry** with **industrial chemistry** becomes particularly evident when examining colorants and preservatives, where compounds serve dual purposes in both food and non-food applications. Understanding these connections provides insight into the broader applications of chemical principles across multiple industries.

The contemporary food industry faces unprecedented challenges in meeting global demand while ensuring safety and sustainability. **Chemical analysis** serves as the cornerstone for regulatory compliance, quality assurance, and product development. The methods and principles discussed in this chapter form the foundation for advanced food science applications and provide the analytical framework necessary for addressing emerging challenges in food security and safety.

1.1 Composition and Analysis of Milk and Dairy Products

Chemical Constituents of Milk

Milk represents one of nature's most complex chemical systems, containing over **100,000 different molecular species** in a delicate colloidal suspension. The primary constituents include proteins, lipids, carbohydrates, minerals, vitamins, and water, each contributing to the nutritional value and functional properties of dairy products. Understanding the **chemical composition** of milk is essential for quality control, processing optimization, and nutritional assessment.

The **protein fraction** of milk consists primarily of caseins (approximately 80%) and whey proteins (20%). Caseins exist as **micelles** with diameters ranging from 50-300 nanometers, stabilized by calcium phosphate bridges and hydrophobic interactions. The four main casein types - α_{s1}-casein, α_{s2}-casein, β-casein, and κ-casein - each possess distinct chemical properties and functional characteristics. The **amphiphilic nature** of caseins, with both hydrophilic and hydrophobic regions, enables their unique ability to form stable micelles in aqueous solution.

Whey proteins include β-lactoglobulin, α-lactalbumin, immunoglobulins, and lactoferrin, each serving specific biological functions. β-lactoglobulin, comprising approximately 50% of whey proteins, contains **free sulfhydryl groups** that contribute to its thermal sensitivity and tendency to form gels upon heating. The presence of disulfide bonds in these proteins significantly influences their **denaturation behavior** and functional properties in processed dairy products.

The **lipid component** of milk exists as an emulsion of fat globules ranging from 0.1 to 15 micrometers in diameter, surrounded by a complex **milk fat globule membrane** (MFGM). This membrane consists of phospholipids, proteins, and glycoproteins that provide stability to the emulsion and influence the **creaming behavior** of milk. The fatty acid composition includes both saturated and unsaturated fatty acids, with oleic acid, palmitic acid, and stearic acid being the most abundant.

Lactose, the primary carbohydrate in milk, exists in both α and β anomeric forms, with the β-form being more soluble and sweeter. The **mutarotation equilibrium** between these forms affects the crystallization behavior of dairy products and influences processing parameters in concentrated dairy applications.

Estimation of Added Water Using Freezing Point Depression

The **freezing point depression** method represents one of the most reliable techniques for detecting water adulteration in milk. This method is based on the **colligative property** that the freezing point of a solution decreases proportionally to the concentration of dissolved particles. Pure milk typically exhibits a freezing point between -0.512°C and -0.536°C, with variations depending on the concentration of dissolved solutes.

The theoretical foundation for this method relies on the relationship:

$$\Delta Tf = Kf \times m \tag{1.1}$$

where ΔTf represents the freezing point depression, Kf is the cryoscopic constant for water (1.86°C·kg/mol), and m is the molality of the solution.

The **Hortvetcryoscope** and **thermistor cryoscope** are commonly employed instruments for precise freezing point determination. The procedure involves cooling the milk sample to approximately -0.6°C while stirring continuously, then allowing controlled crystallization to occur. The temperature at which the first ice crystals form represents the **true freezing point** of the sample.

Added water detection becomes possible because dilution reduces the concentration of dissolved solutes, thereby raising the freezing point toward 0°C. Each 1% addition of water typically raises the freezing point by approximately **0.0055°C**. This relationship enables quantitative determination of adulteration levels with high precision.

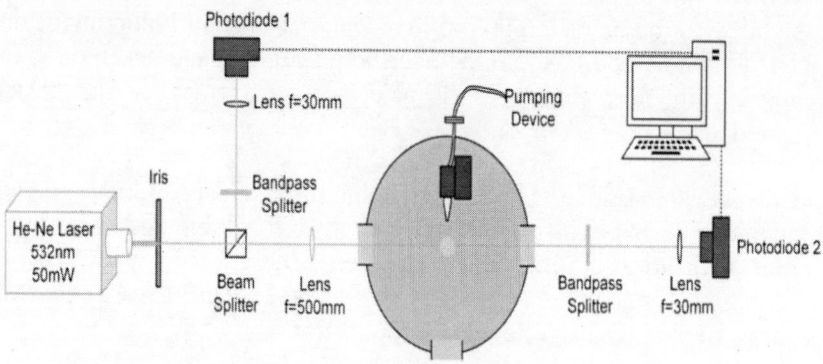

Fig. 1: Schematic representation of a thermistor cryoscope showing the cooling chamber, stirrer, and temperature measurement system

Fat Content Analysis Using Gerber and Babcock Methods

Fat content determination remains one of the most critical quality parameters for dairy products, influencing both nutritional labeling and economic value. The **Gerber method** and **Babcock method** represent classical approaches that continue to serve as reference methods despite the availability of modern instrumental techniques.

The **Gerber method** utilizes concentrated sulfuric acid to digest the protein matrix and release fat globules, followed by centrifugation to separate the fat layer. The procedure involves mixing 10 mL of milk with 10 mL of concentrated sulfuric acid (density 1.82 g/mL) and 1 mL of isoamyl alcohol in a specialized **Gerber butyrometer**. The sulfuric acid serves multiple functions: it dissolves proteins through **acid**

hydrolysis, generates heat through the exothermic reaction, and increases the density of the aqueous phase to facilitate fat separation.

The chemical reactions involved include protein denaturation and partial hydrolysis:

$$\text{Protein} + H_2SO_4 \rightarrow \text{Hydrolyzed peptides} + \text{Heat} \quad (1.2)$$

The **isoamyl alcohol** prevents charring of organic matter and aids in the complete separation of fat by reducing surface tension. After centrifugation at 1,100 rpm for 5 minutes, the fat column is read directly from the graduated neck of the butyrometer.

The **Babcock method** follows similar principles but employs a different acid concentration and includes additional procedural modifications. This method uses sulfuric acid with a specific gravity of 1.82-1.83 and requires precise temperature control during the digestion process. The **sulfuric acid concentration** is critical, as insufficient acid strength results in incomplete protein digestion, while excessive concentration can char the fat and produce erroneous results.

Both methods demonstrate excellent **repeatability and reproducibility** when performed correctly, with coefficients of variation typically less than 2%. The accuracy of these methods depends on proper calibration of the butyrometers, standardization of reagent concentrations, and adherence to specified centrifugation parameters.

Mineral Profiling Using Atomic Absorption Spectroscopy

Atomic absorption spectroscopy (AAS) has become the method of choice for **mineral analysis** in dairy products due to its high sensitivity, selectivity, and relative freedom from matrix interferences. The technique is particularly valuable for determining essential minerals such as calcium, magnesium, iron, zinc, and copper, as well as potentially toxic elements like lead and cadmium.

The fundamental principle of AAS involves the **atomization** of sample elements in a flame or graphite furnace, followed by measurement of light absorption at element-specific wavelengths. For dairy applications, **flame atomization** using an air-acetylene flame is commonly employed for major and minor elements, while **graphite furnace** atomization provides enhanced sensitivity for trace elements.

Sample preparation for dairy analysis typically involves **acid digestion** to destroy the organic matrix and solubilize minerals. A typical digestion procedure involves treating 5-10 grams of sample with a mixture of nitric acid and hydrogen peroxide, followed by heating to complete oxidation of organic matter:

$$\text{Organic matter} + HNO_3 + H_2O_2 \rightarrow CO_2 + H_2O + NO_2 + \text{minerals} \quad (1.3)$$

Matrix effects in dairy samples can significantly influence analytical results, particularly for elements like calcium and magnesium that are present at high concentrations. The high **protein content** and **fat content** can cause both physical and chemical interferences. Physical interferences include changes in surface tension and viscosity that affect nebulization efficiency, while chemical interferences involve the formation of refractory compounds that resist atomization.

Ionization interference becomes particularly significant for alkali metals like sodium and potassium in the presence of easily ionized elements. This effect can be minimized through the addition of **ionization suppressors** such as cesium or lanthanum chloride. Similarly, **chemical interference** from phosphate ions forming stable compounds with calcium can be overcome using **releasing agents** like strontium or lanthanum.

The **method of standard additions** is frequently employed to compensate for matrix effects, involving the addition of known amounts of analyte to sample aliquots. This approach allows for the construction of calibration curves in the presence of the sample matrix, thereby accounting for any systematic interferences.

Quality control measures include the use of certified reference materials (CRMs) such as NIST SRM 1549a (Non-Fat Milk Powder) and regular participation in proficiency testing programs. The precision of AAS methods for major elements typically achieves relative standard deviations of 2-5%, while trace elements may exhibit somewhat higher variability depending on concentration levels.

1.2 Beverage Chemistry and Additive Analysis

Estimation of Caffeine via HPLC

High-Performance Liquid Chromatography (HPLC) has emerged as the definitive method for **caffeine quantification**in beverages due to its exceptional specificity, sensitivity, and ability to separate caffeine from structurally similar compounds. The method's importance extends beyond simple quantification to include the analysis of caffeine metabolites and related methylxanthines such as **theophylline** and **theobromine**.

The chemical structure of **caffeine** (1,3,7-trimethylxanthine) provides unique spectroscopic properties that facilitate its detection and quantification. The compound exhibits strong **UV absorption** with maximum absorbance at 273 nm, making UV detection the most common approach for HPLC analysis. The presence of the **purine ring system** contributes to caffeine's stability under normal analytical conditions while providing the chromophore necessary for sensitive detection.

Sample preparation for caffeine analysis varies depending on the beverage matrix but generally involves **extraction** and **cleanup** steps to remove interfering compounds. For coffee and tea samples, **liquid-liquid extraction** using dichloromethane or chloroform effectively isolates caffeine from tannins and other polar compounds. Carbonated beverages typically require only **degassing** and dilution, while alcoholic beverages may necessitate **deproteinization**steps.

The most widely employed HPLC method utilizes a **reverse-phase C18 column** with a mobile phase consisting of water and acetonitrile or methanol. A typical gradient program begins with 95% aqueous phase, decreasing to 70% over 20 minutes to ensure complete elution of caffeine and related compounds. The **buffer system** commonly employs phosphate buffer (pH 3.5) to maintain consistent retention times and peak shapes.

Method validation for caffeine analysis demonstrates excellent linearity over the concentration range of 1-500 mg/L, with correlation coefficients typically exceeding 0.999. The **limit of detection** (LOD) and **limit of quantification** (LOQ) are approximately 0.1 mg/L and 0.3 mg/L, respectively, when using UV detection at 273 nm. **Recovery**

studies using spiked samples consistently yield values between 95-105%, confirming the method's accuracy.

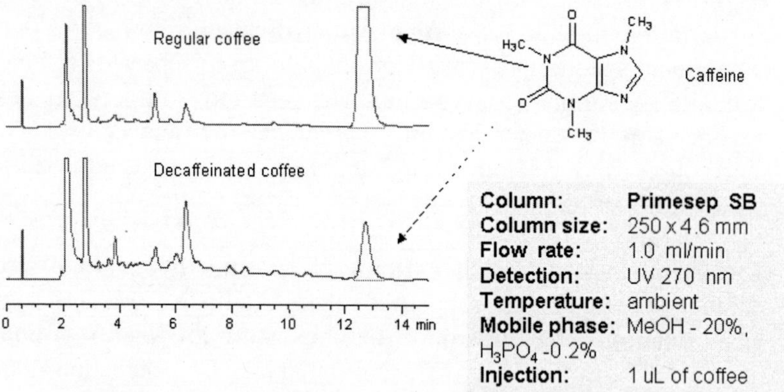

Fig. 2: Representative HPLC chromatogram showing caffeine separation from related methylxanthines in coffee extract

Detection of Chicory Using TLC Methods

The **adulteration of coffee** with chicory represents a significant economic and quality concern in the global coffee industry. **Thin Layer Chromatography** (TLC) provides a cost-effective and reliable method for detecting chicory adulteration, based on the distinct chemical profiles of **roasted chicory root** compared to **coffee beans.**

Chicory (Cichorium intybus) contains characteristic compounds including **inulin**, **chicoric acid**, and various **sesquiterpene lactones** that are absent in pure coffee. The roasting process creates additional **Maillard reaction products** and **pyrolysis compounds** that provide unique fingerprint patterns for identification.

The analytical procedure begins with **solvent extraction** using a mixture of methanol and water (70:30 v/v) to extract both polar and moderately polar compounds. The extraction is typically performed using **Soxhlet extraction** or **ultrasonic-assisted extraction** to ensure complete recovery of target compounds. The extract is then concentrated and spotted onto **silica gel TLC plates** using calibrated micropipettes.

The **mobile phase** selection is critical for achieving optimal separation of chicory markers from coffee components. A commonly employed system consists of **ethyl acetate:methanol:water** (100:13.5:10 v/v/v), which provides excellent resolution of the key discriminating compounds. Alternative mobile phases include **chloroform:methanol:water**(65:35:10 v/v/v) for enhanced separation of lipophilic components.

Visualization of the developed chromatograms employs multiple detection methods to ensure comprehensive analysis. **UV examination** at 254 nm and 366 nm reveals compounds with aromatic chromophores, while **vanillin-sulfuric acid reagent** produces characteristic color reactions with different compound classes. The reagent preparation involves dissolving 1% vanillin in concentrated sulfuric acid, followed by heating at 110°C for 5-10 minutes.

Chicoric acid appears as a prominent spot with an Rf value of approximately 0.35-0.40 in the ethyl acetate system, producing a **blue-green fluorescence** under UV light and a **reddish-brown color** with vanillin reagent. **Inulin degradation products** migrate to different positions, creating a characteristic pattern that distinguishes chicory from pure coffee extracts.

Quantitative analysis can be achieved through **densitometric scanning** of the developed plates, measuring peak areas at specific wavelengths. The method demonstrates good linearity for chicory concentrations ranging from 5-50% in coffee blends, with detection limits of approximately 2-3% chicory content.

Analysis of Methyl Alcohol Using Dichromate Oxidation

Methanol detection in alcoholic beverages represents a critical safety concern due to the compound's **toxic metabolites** - formaldehyde and formic acid - which can cause blindness and death. The **dichromate oxidation method** provides a reliable, cost-effective approach for methanol quantification based on the **stoichiometric oxidation** of methanol to formaldehyde and subsequent colorimetric determination.

The chemical principle involves the **reduction of dichromate ions** by methanol in acidic solution, producing **chromic ions** and **formaldehyde**:

$$3CH_3OH + 2K_2Cr_2O_7 + 8H_2SO_4 \rightarrow 3CH_2O + 2K_2SO_4 + 2Cr_2(SO_4)_3 + 11H_2O \quad (1.4)$$

The **formaldehyde product** subsequently reacts with **chromotropic acid** (1,8-dihydroxynaphthalene-3,6-disulfonic acid) to form a **violet-colored complex** with maximum absorption at 575 nm. This colorimetric reaction demonstrates high specificity for formaldehyde, minimizing interference from other alcohols and organic compounds.

Sample preparation involves **distillation** to separate methanol from the complex beverage matrix while removing potential interferents. The distillation process typically employs a **simple distillation apparatus** with careful temperature control to prevent co-distillation of higher alcohols. The first 10-15% of the distillate contains the majority of methanol and is collected for analysis.

The analytical procedure requires **precise pH control** and **temperature regulation** to ensure complete oxidation and consistent color development. The reaction mixture contains the distilled sample, **potassium dichromate solution** (5% w/v), and **concentrated sulfuric acid** in specific proportions. The mixture is heated at 60°C for 15 minutes to ensure complete oxidation, followed by cooling and addition of **chromotropic acid reagent**.

Interference studies have identified several compounds that can affect the analysis, including **higher alcohols**, **aldehydes**, and **organic acids**. **Ethanol**, the primary alcohol in beverages, can interfere at high concentrations by competing for oxidation sites. This interference is minimized through the use of **controlled reaction conditions** and appropriate **dilution factors**.

Method validation demonstrates excellent linearity over the range of 0.1-10 mg/L methanol, with a correlation coefficient of 0.998. The **detection limit** is approximately 0.05 mg/L, while the **quantification limit** is 0.15 mg/L. **Recovery studies** using spiked wine and spirit samples yield values between 92-108%, confirming the method's accuracy for routine analysis.

Identification of Chloral Hydrate in Toddy

Chloral hydrate (2,2,2-trichloroethanediol) detection in **toddy** and other traditional fermented beverages addresses both safety and regulatory concerns. This compound, historically known as **"knockout drops"** or **"Mickey Finn,"** poses serious health risks due to its **sedative properties** and potential for **respiratory depression**.

The **chemical properties** of chloral hydrate include high water solubility, volatility, and characteristic **halogenated structure** that provides unique analytical signatures. The compound exists in equilibrium between the **hydrated form**(chloral hydrate) and the **anhydrous form** (trichloroacetaldehyde), with the hydrated form predominating in aqueous solution:

$$CCl_3CHO + H_2O \rightleftharpoons CCl_3CH(OH)_2 \qquad (1.5)$$

Analytical approaches for chloral hydrate detection employ **gas chromatography** with **electron capture detection**(GC-ECD) or **mass spectrometry** (GC-MS) due to the compound's volatility and the high sensitivity of halogenated compounds to these detection methods. The **electron capture detector** exhibits exceptional sensitivity to halogen-containing compounds, achieving detection limits in the **parts-per-billion range**.

Sample preparation involves **liquid-liquid extraction** using organic solvents such as **diethyl ether** or **ethyl acetate** to isolate chloral hydrate from the complex toddy matrix. The **pH adjustment** to slightly acidic conditions (pH 2-3) enhances extraction efficiency by maintaining chloral hydrate in its **molecular form**. **Salting out** with sodium chloride further improves extraction recovery by reducing the compound's solubility in the aqueous phase.

Derivatization reactions can enhance the analytical performance by creating more stable, easily detectable derivatives. **Pentafluorobenzyl hydroxylamine** (PFBHA) reacts with the aldehyde group to form **oxime derivatives** that exhibit improved chromatographic behavior and enhanced mass spectral fragmentation patterns.

Quality control measures include the use of **deuterated internal standards** such as chloral hydrate-d_1, which compensate for variations in extraction efficiency and instrumental response. **Method blanks** and **procedural blanks** are essential to identify potential contamination sources, while **positive controls** using spiked samples verify the method's capability to detect chloral hydrate at relevant concentrations.

The **regulatory implications** of chloral hydrate detection extend beyond immediate safety concerns to include **food adulteration laws** and **public health protection**. Many jurisdictions classify chloral hydrate as a **controlled substance**, making its presence in beverages a serious legal violation requiring immediate regulatory action.

1.3 Food Preservatives, Additives, and Adulterants

Chemical Nature of Benzoates, Sorbates, and Propionates

Food preservation through chemical additives represents one of the most significant advances in food technology, enabling extended shelf life while maintaining **nutritional quality** and **sensory properties**. The three major classes of **organic acid preservatives** - benzoates, sorbates, and propionates - each possess distinct **antimicrobial mechanisms** and **application characteristics** that make them suitable for different food systems..

Benzoic acid and its salts function primarily through **membrane disruption** and **enzyme inhibition** in microbial cells. The **undissociated acid form** is the active species, requiring pH levels below the **pKa value** (4.19) for optimal efficacy. The **lipophilic nature** of undissociated benzoic acid enables penetration through microbial cell membranes, where it disrupts **proton-motive force** and interferes with **enzymatic processes**. The **Henderson-Hasselbalch equation** describes the relationship between pH and the proportion of undissociated acid:

$$pH = pKa + \log([A^-]/[HA]) \quad (1.6)$$

Sodium benzoate represents the most commonly used form due to its excellent **water solubility** (556 g/L at 20°C) compared to benzoic acid (2.9 g/L at 20°C). The compound demonstrates **broad-spectrum activity** against yeasts and molds but exhibits limited effectiveness against bacteria, particularly **gram-positive species**. The **maximum permitted levels** typically range from 0.1% to 0.2% depending on the food category and regulatory jurisdiction.

Sorbic acid and its potassium salt exhibit **superior antimicrobial activity** compared to benzoates, particularly against molds and yeasts. The compound's mechanism involves **interaction with sulfhydryl groups** in microbial enzymes, disrupting cellular metabolism. The **conjugated double bond system** in sorbic acid contributes to its antimicrobial properties while also making it susceptible to **oxidative degradation** and **photochemical reactions**.

The **chemical stability** of sorbates varies significantly with environmental conditions. **Heat processing** can lead to decarboxylation reactions, while **alkaline conditions** promote hydrolysis. The presence of **metal ions**, particularly iron and copper, catalyzes oxidative degradation through **free radical mechanisms**. These factors necessitate careful consideration of processing conditions and storage environments when employing sorbate preservatives.

Propionic acid and its salts demonstrate **selective antimicrobial activity**, being particularly effective against **molds and rope-forming bacteria** while having minimal impact on **lactic acid bacteria**. This selectivity makes propionates ideal for **bakery products** where maintaining **yeast activity** is essential during fermentation while preventing **post-baking contamination**. The compound's mechanism involves **disruption of cellular metabolism** through interference with **tricarboxylic acid cycle** enzymes.

Synergistic effects between different preservative systems can enhance overall antimicrobial efficacy while reducing individual compound concentrations. The combination of **benzoates with sorbates** often produces additive or synergistic antimicrobial effects, particularly in **acidic food systems**. These interactions enable formulators to achieve desired preservation levels while minimizing potential **sensory impacts** and **regulatory concerns**.

Analysis of Artificial Sweeteners: Saccharin, Aspartame, Sucralose

The **analytical chemistry** of artificial sweeteners presents unique challenges due to their **structural diversity**, **varying physicochemical properties**, and **complex food matrices**. **Modern analytical methods** must address regulatory requirements for accurate quantification while dealing with potential **matrix interferences** and **degradation products**.

Saccharin (1,2-benzisothiazol-3(2H)-one 1,1-dioxide) represents the oldest synthetic sweetener, possessing a **distinctive chemical structure** with both **acidic and basic properties**. The compound exhibits **amphoteric behavior** with a pKa of 1.6, making it predominantly ionized under normal food pH conditions. This **ionic character**

influences both its **extraction behavior** and **chromatographic retention** in analytical methods.

HPLC analysis of saccharin typically employs **reverse-phase chromatography** with **UV detection** at 230 nm, where the compound exhibits strong absorption due to its **aromatic ring system**. The **mobile phase** commonly consists of **phosphate buffer** (pH 3.0) and acetonitrile in varying proportions, optimized to achieve baseline separation from potential interferents including **cyclamate** and **acesulfame-K**.

Sample preparation for saccharin analysis involves **protein precipitation** using acetonitrile or methanol, followed by **centrifugation** and **filtration**. The **Carrez clarification** procedure using **zinc acetate** and **potassium ferrocyanide**effectively removes proteins and other macromolecular interferents while maintaining saccharin stability.

Aspartame (N-L-α-aspartyl-L-phenylalanine 1-methyl ester) presents unique analytical challenges due to its **inherent instability** under certain pH and temperature conditions. The compound undergoes **hydrolysis** to form **aspartic acid**, **phenylalanine**, and **methanol**, making **sample handling** and **storage conditions** critical factors in analytical accuracy.

$$\text{Aspartame} + H_2O \rightarrow \text{Aspartic acid} + \text{Phenylalanine} + \text{Methanol} \quad (1.7)$$

The **degradation kinetics** of aspartame follow **first-order kinetics**, with the rate constant being dependent on **temperature**, **pH**, and **ionic strength**. Maximum stability occurs at **pH 4.3**, while acidic and alkaline conditions accelerate degradation. These characteristics necessitate **rapid sample processing** and **analysis** to prevent significant losses during analytical procedures.

LC-MS/MS methods have become increasingly popular for aspartame analysis due to their **enhanced specificity** and ability to **monitor degradation products** simultaneously. The **electrospray ionization** (ESI) interface provides excellent sensitivity for aspartame and its metabolites, enabling detection limits below 1 mg/L in complex food matrices.

Sucralose (4,1',6'-trichloro-4,1',6'-trideoxygalactosucrose) demonstrates exceptional **chemical stability** compared to other artificial sweeteners, remaining **unchanged** under normal food processing conditions. This stability advantage simplifies analytical procedures but presents challenges in **sample preparation** due to the compound's **high polarity** and **limited extractability** with conventional organic solvents.

Analytical methods for sucralose typically employ **hydrophilic interaction chromatography** (HILIC) or **ion-pair chromatography** to achieve adequate retention and separation. The compound lacks **UV-absorbing chromophores**, necessitating the use of **evaporative light scattering detection** (ELSD) or **mass spectrometry** for sensitive quantification.

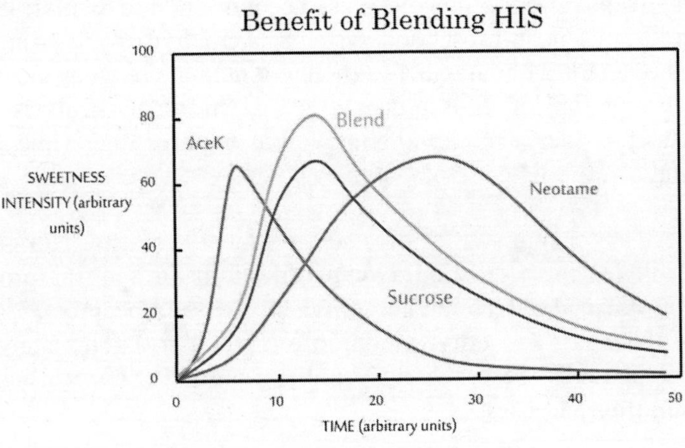

Fig. 3: Comparative stability profiles of artificial sweeteners under different pH and temperature conditions

Testing Protocols for MSG and Fruit Esters

Monosodium glutamate (MSG) analysis requires **specific methodological approaches** due to the compound's **amino acid nature** and **ubiquitous presence** in protein-containing foods. The distinction between **added MSG** and **naturally occurring glutamate** from protein hydrolysis presents significant analytical challenges that require sophisticated **separation techniques** and **quantification strategies**.

The **chemical properties** of glutamate include **zwitterionic behavior** at physiological pH, **strong water solubility**, and **lack of UV absorption**, making conventional **reverse-phase HPLC** unsuitable for direct analysis. **Ion-exchange chromatography** or **derivatization methods** are typically employed to enable adequate separation and detection.

Pre-column derivatization using **o-phthaldialdehyde** (OPA) or **9-fluorenylmethyloxycarbonyl chloride** (FMOC-Cl) creates **fluorescent derivatives** that enable sensitive detection using **fluorescence detection**. The OPA derivatization reaction occurs rapidly at **room temperature** in alkaline conditions:

Glutamate + OPA + RSH → Fluorescent isoindole derivative (1.8)

Sample preparation protocols must account for the **protein-bound glutamate** present in most food systems. **Acid hydrolysis** using 6 M HCl at 110°C for 24 hours releases bound glutamate but may also cause some destruction of free glutamate. **Enzymatic hydrolysis** using **proteases** provides a milder alternative but requires longer incubation times and careful **pH control**.

Fruit esters encompass a diverse group of **volatile compounds** responsible for characteristic **flavor profiles** in fruits and fruit products. Common **natural esters** include **ethyl butyrate** (pineapple), **isoamyl acetate** (banana), and **ethyl cinnamate** (strawberry). The analysis of these compounds serves both **quality control** and **authenticity determination**purposes.

Gas chromatographic analysis represents the method of choice for fruit ester determination due to the **volatile nature** of these compounds and the excellent **separation capabilities** of GC systems. **Capillary columns** with **polar stationary phases** such as **polyethylene glycol** provide optimal separation of structurally similar esters.

Sample preparation typically involves **headspace analysis** or **solid-phase microextraction** (SPME) to concentrate volatile esters while minimizing **matrix interferences**. **Dynamic headspace** techniques using **purge-and-trap** systems achieve excellent sensitivity and reproducibility for volatile ester analysis.

Static headspace analysis offers simplicity and reproducibility, with **equilibrium conditions** established at elevated temperatures (typically

60-80°C) to enhance **vapor phase concentrations**. The **partition coefficients** between liquid and vapor phases determine the **sensitivity** and **detection limits** achievable with this approach.

Mass spectrometric detection provides **definitive identification** of fruit esters through **characteristic fragmentation patterns**. **Electron impact ionization** produces predictable fragmentation, with **molecular ion peaks** and **base peaks**corresponding to **loss of alkoxy groups** or **acyl groups** depending on the ester structure.

1.4 Colorants, Contaminants, and Pesticide Residue

Detection of Coal Tar Dyes and Non-Permitted Colors

The analysis of **synthetic food colorants** represents a critical aspect of food safety and regulatory compliance, addressing both **permitted artificial colors** and **prohibited substances** that may be present due to **adulteration** or **cross-contamination**. **Coal tar dyes**, derived from **petroleum-based aromatics**, constitute the majority of synthetic food colorants but include both **approved** and **banned substances** with significantly different **toxicological profiles**.

Coal tar dyes are characterized by their **azo**, **triphenylmethane**, **xanthene**, or **quinoline** chromophore systems, each providing distinct **spectroscopic properties** and **chemical reactivity**. The **azo dyes**, containing one or more -N=N- linkages, represent the largest class and include both **permitted colors** such as **Allura Red AC** and **prohibited substances** such as **Sudan dyes**. The **electronic transitions** in these compounds produce intense colors in the **visible spectrum**, making **spectrophotometric detection** highly sensitive.

Regulatory frameworks worldwide maintain **positive lists** of permitted food colorants with specific **purity criteria** and **maximum usage levels**. **Non-permitted colors** include industrial dyes, **banned food colorants**, and **impurities** that may pose health risks. The detection of these substances requires **comprehensive analytical methods** capable of **simultaneous analysis** of multiple compounds with varying **physicochemical properties**.

HPLC-DAD analysis (High-Performance Liquid Chromatography with Diode Array Detection) has emerged as the standard method for **synthetic colorant analysis** due to its ability to provide both

quantitative determination and **spectral identification**. The **diode array detector** records complete **UV-visible spectra** for each chromatographic peak, enabling **peak purity assessment** and **compound identification** through **spectral library matching**.

Sample preparation involves **extraction** of colorants from the food matrix using appropriate solvents, typically **aqueous solutions** at **controlled pH** to ensure complete recovery of both **acidic** and **basic dyes**. **Solid-phase extraction** (SPE) using **C18** or **polyamide cartridges** provides effective cleanup and concentration, removing **matrix interferents** while retaining target analytes.

The **chromatographic separation** employs **gradient elution** with **aqueous buffers** and **organic modifiers** to achieve baseline resolution of structurally similar compounds. A typical mobile phase system consists of **ammonium acetate buffer** (pH 6.8) and **acetonitrile**, with the gradient programmed to elute **hydrophilic dyes** first, followed by **lipophilic compounds**. The **buffer pH** is critical as it affects the **ionization state** of acidic and basic dyes, directly influencing their **retention behavior** and **peak symmetry**.

Method validation for colorant analysis must demonstrate **selectivity** against a comprehensive range of **permitted and non-permitted dyes**. **Spectral overlap** between structurally related compounds requires careful **wavelength selection**and**chromatographic optimization**. The **detection wavelengths** are typically selected based on the **absorption maxima**of target compounds, commonly including 254 nm, 430 nm, 520 nm, and 630 nm to cover the major chromophore classes.

Sudan dye detection requires particular attention due to their **lipophilic nature** and **potential presence** in spice-based food products. These **illegal colorants** exhibit **poor water solubility** but demonstrate excellent **fat solubility**, necessitating **alternative extraction procedures** using **organic solvents** such as **acetone** or **ethyl acetate**. The **toxicological concerns** associated with Sudan dyes, including **genotoxicity** and **carcinogenicity**, make their detection a high priority in **food surveillance programs**.

Confirmation methods using **LC-MS/MS** provide **unequivocal identification** through **molecular ion detection** and **characteristic fragmentation patterns**. The **electrospray ionization** interface operates in both **positive** and **negative modes** to accommodate the diverse **chemical properties** of different dye classes. **Multiple reaction monitoring** (MRM) enables **simultaneous quantification** of numerous compounds with **enhanced selectivity** and **reduced matrix effects.**

Heavy Metal Analysis in Food Using ICP-MS

Inductively Coupled Plasma Mass Spectrometry (ICP-MS) represents the **gold standard** for **heavy metal analysis** in food systems, providing **multi-element capability, exceptional sensitivity**, and **wide linear dynamic range**. The technique addresses growing concerns about **toxic metal contamination** in the global food supply, including **lead, cadmium, mercury**, and **arsenic**, which pose significant **public health risks** even at trace concentrations.

The **fundamental principles** of ICP-MS involve **sample nebulization, aerosol transport, plasma ionization**, and **mass spectrometric detection**. The **inductively coupled plasma** operates at temperatures exceeding **6000 K**, ensuring complete **atomization** and **ionization** of most elements. The **high temperature** and **chemically inert environment**minimize**chemical interferences** while providing **robust ionization** across the periodic table.

Sample preparation for food analysis typically requires **complete digestion** of the organic matrix to prevent **carbon deposition** on the plasma torch and **interface cones. Microwave-assisted acid digestion** using **nitric acid** and **hydrogen peroxide** provides **rapid, complete,** and **reproducible** sample decomposition while minimizing **volatile element losses**. The digestion procedure follows a controlled **temperature program**:

$$\text{Food matrix} + HNO_3 + H_2O_2 \rightarrow CO_2 + H_2O + NO_2 + \text{dissolved metals} \quad (1.9)$$

Spectral interferences in ICP-MS arise from **isobaric overlaps, polyatomic ion formation**, and **doubly charged species. Polyatomic interferences** are particularly problematic for elements like **arsenic** ($^{75}As^{+}$ vs $^{40}Ar^{35}Cl^{+}$) and **selenium** ($^{80}Se^{+}$ vs $^{40}Ar^{2+}$). **Collision/reaction cell technology** effectively reduces these interferences through **gas-phase reactions** that selectively eliminate **polyatomic ions** while preserving **analyte signals**.

Matrix effects in food samples can significantly affect **ionization efficiency** and **signal intensity**. **High dissolved solid content** from digested samples can cause **signal suppression** or **enhancement** depending on the element and matrix composition. **Internal standardization** using elements such as **scandium**, **rhodium**, and **bismuth** compensates for these effects by correcting for **instrumental drift** and **matrix-induced signal changes**.

Quality assurance measures include the use of **certified reference materials** (CRMs) such as NIST SRM 1568b (Rice Flour) and IRMM-804 (Rice). These materials provide **traceability** to international standards and enable **method validation** through comparison of measured values with **certified concentrations**. **Detection limits** for most toxic metals range from **0.1 to 1 μg/kg**, well below **regulatory action levels** established by **international food safety authorities**.

Data interpretation requires consideration of **measurement uncertainty**, **analytical bias**, and **biological variability**. **Expanded uncertainty** calculations incorporate contributions from **sampling**, **sample preparation**, **instrumental analysis**, and **calibration standards**. The **fitness for purpose** of analytical results depends on their **uncertainty** relative to **regulatory limits** and **health-based guidance values**.

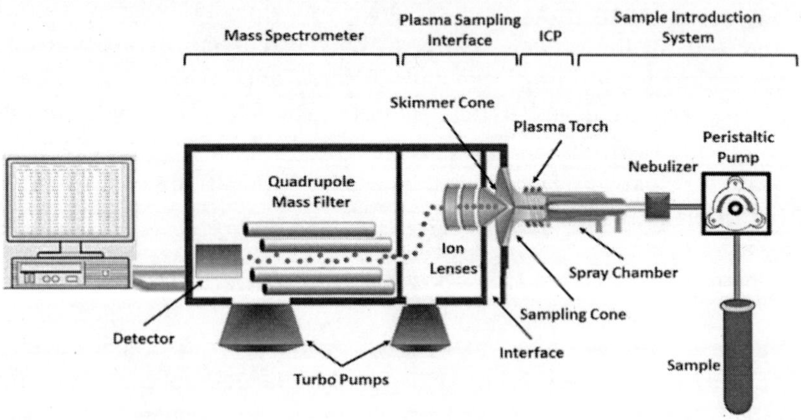

Fig. 4: Schematic diagram of ICP-MS system showing sample introduction, plasma torch, interface region, and mass analyzer components

GC-MS Analysis for Pesticide Residues

Gas Chromatography-Mass Spectrometry (GC-MS) has established itself as the **premier analytical technique** for **pesticide residue**

analysis due to its exceptional **separation efficiency**, **sensitive detection**, and **reliable identification** capabilities. The method addresses **food safety concerns** related to **agricultural chemical residues** while supporting **regulatory compliance** and **international trade requirements.**

Pesticide chemistry encompasses diverse structural classes including **organochlorines**, **organophosphates**, **carbamates**, **pyrethroids**, and **neonicotinoids**, each presenting unique **analytical challenges**. The **physicochemical properties** of these compounds - including **volatility**, **thermal stability**, and **polarity** - directly influence their **GC behavior** and **detection characteristics**.

Sample preparation methodologies must accommodate the **wide polarity range** of pesticide residues while removing **matrix interferents** that can compromise **analytical performance**. The **QuEChERS method** (Quick, Easy, Cheap, Effective, Rugged, and Safe) has gained widespread acceptance for **multi-residue analysis,** providing **high recovery** and **good precision** across diverse **food matrices.**

The **QuEChERS procedure** involves **initial extraction** using **acetonitrile** in the presence of **anhydrous magnesium sulfate** and **sodium chloride**, followed by **dispersive solid-phase extraction** (d-SPE) cleanup. The **salting-out effect** enhances **phase separation** while the **magnesium sulfate** removes **residual water** that could interfere with **GC analysis**:

Sample + CH_3CN + $MgSO_4$ + NaCl → Extracted pesticides + salted aqueous phase (1.10)

Cleanup procedures using **primary secondary amine** (PSA) and **C18** sorbents remove **polar matrix components** and **lipids**, respectively. **Graphitized carbon black** (GCB) effectively removes **pigments** and **sterols** but requires careful optimization to prevent **planar pesticide losses.**

Gas chromatographic separation employs **capillary columns** with **low-polarity stationary phases** such as **5% phenyl methylpolysiloxane** to achieve optimal **resolution** of pesticide mixtures. **Temperature programming** typically begins at **70-80°C**

with **rapid heating rates** (15-25°C/min) to minimize **analysis time** while maintaining **separation quality**.

Mass spectrometric detection operates in **electron impact** (EI) mode at **70 eV**, producing **characteristic fragmentation patterns** for compound identification. **Selected Ion Monitoring** (SIM) enhances **sensitivity** by monitoring specific **m/z values** for each target compound, while **full-scan acquisition** provides **comprehensive spectral information** for **unknown compound identification**.

Method validation follows international guidelines such as **SANTE/12682/2019**, requiring demonstration of **linearity**, **accuracy**, **precision**, **selectivity**, and **stability**. **Recovery studies** using **blank samples** spiked at **0.01**, **0.1**, and **1.0 mg/kg**must yield values between **70-120%** with **relative standard deviations** below **20%**.

Quality control measures include **concurrent analysis** of **matrix blanks**, **solvent blanks**, and **positive controls** with each **analytical batch**. **Proficiency testing** participation through programs such as **FAPAS** provides **external validation**of analytical performance and enables **interlaboratory comparisons**.

Data interpretation requires consideration of **maximum residue limits** (MRLs) established by **regulatory authorities**such as the **Codex Alimentarius**, **EPA**, and **European Commission**. **Exceedances** of established limits trigger **risk assessment** procedures and may result in **regulatory action** including **product recalls** or **import restrictions**.

1.5 Paints, Pigments, and Industrial Dyes

Chemical Classification of Pigments: TiO_2, ZnO, Lithopone

Inorganic pigments constitute the foundation of **modern coating systems**, providing essential properties including **color**, **opacity**, **durability**, and **protection** against **environmental degradation**. The three major **white pigments** - **titanium dioxide**, **zinc oxide**, and **lithopone** - demonstrate distinct **chemical characteristics** and **performance attributes** that determine their **application suitability** in various coating formulations.

Titanium dioxide (TiO_2) exists in three **crystalline forms**: **rutile**, **anatase**, and **brookite**, with **rutile** and **anatase** being the commercially significant polymorphs. **Rutile TiO_2** exhibits superior **weather resistance** and **chalking resistance** due to its **thermodynamically stable crystal structure**, making it the preferred choice for **exterior applications**. The **refractive index** of rutile (2.76) provides exceptional **hiding power** and **brightness**, while **anatase** (refractive index 2.55) offers enhanced **photocatalytic activity** but reduced **durability** under **UV exposure.**

The **particle size distribution** of TiO_2 pigments significantly influences **optical properties** and **coating performance. Optimal light scattering** occurs when particle sizes approximate **half the wavelength** of visible light (200-300 nm), maximizing **opacity** and **tinting strength. Surface treatments** using **aluminum oxide, silicon dioxide**, or **organic compounds** modify **surface chemistry** to improve **dispersibility, reduce photocatalytic activity**, and enhance **coating durability**.

Zinc oxide (ZnO) provides unique **multifunctional properties** including **UV absorption**, **antimicrobial activity**, and **mildew resistance**. The compound's **wide bandgap** (3.37 eV) enables **UV-filtering capability** while maintaining **transparency** in the visible spectrum. **Zinc oxide** demonstrates **amphoteric behavior**, reacting with both **acids** and **bases**, which can influence **coating stability** in extreme pH environments.

The **antimicrobial mechanism** of zinc oxide involves **generation of reactive oxygen species** (ROS) upon **UV irradiation** and **zinc ion release** that disrupts **microbial cell membranes**. These properties make ZnO particularly valuable in **marine coatings**, **architectural paints**, and **antimicrobial applications**. However, the **catalytic activity** can also promote **coating degradation** through **oxidative processes**, requiring careful **formulation balance**.

Lithopone represents a **coprecipitated mixture** of **zinc sulfide** (ZnS) and **barium sulfate** ($BaSO_4$), typically containing 28-30% ZnS by weight. The **precipitation process** involves simultaneous formation of both compounds from **aqueous solutions**:

$$ZnSO_4 + BaS \rightarrow ZnS \cdot BaSO_4 \text{ (lithopone)} + \text{precipitate} \quad (1.11)$$

The **composite structure** provides **intermediate properties** between pure ZnS and $BaSO_4$, offering **good hiding power** with **improved**

lightfastness compared to zinc sulfide alone. **Lithopone** exhibits **excellent color retention** and **alkali resistance** but demonstrates **sensitivity to acid conditions** where zinc sulfide can dissolve.

Particle morphology and **surface characteristics** influence **pigment dispersion** and **coating rheology**. **Spherical particles** typically provide **easier dispersion** and **lower viscosity**, while **acicular** or **platelet structures** may enhance **barrier properties** but increase **dispersion energy requirements**. **Surface area measurements** using **BET analysis** provide critical information for **formulation optimization** and **quality control**.

Functional Requirements for Good Paints: Adhesion, Opacity

Paint performance depends on achieving optimal **balance** among multiple **functional requirements**, with **adhesion** and **opacity** representing two of the most critical properties. **Adhesion** ensures **coating integrity** and **substrate protection**, while **opacity** provides **aesthetic appeal** and **hiding power** essential for commercial acceptance. Understanding the **chemical** and **physical mechanisms** underlying these properties enables **formulation optimization** and **performance prediction**.

Adhesion mechanisms in paint systems involve **mechanical interlocking**, **chemical bonding**, **electrostatic interactions**, and **van der Waals forces**. **Mechanical adhesion** occurs through **penetration** of liquid coating into **substrate porosity** and **surface irregularities**, creating **physical anchoring** upon curing. The **surface roughness** and **profile** significantly influence this mechanism, with **optimal roughness** balancing **increased contact area** against **excessive coating consumption**.

Chemical adhesion involves formation of **covalent bonds**, **hydrogen bonds**, or **coordination complexes** between **coating components** and **substrate surfaces**. **Primer systems** often incorporate **reactive groups** such as **silanes**, **titanates**, or **chromates** that form **chemical bridges** between **inorganic substrates** and **organic coatings**. The **surface energy** of substrates must exceed the **surface tension** of liquid coatings to ensure **proper wetting** and **adhesion development**.

Surface preparation represents a critical factor in **adhesion performance**, involving **contamination removal**, **surface activation**, and **profile generation**. **Chemical cleaning** removes **oils**, **salts**, and **organic contaminants** that can interfere with **adhesion mechanisms**. **Mechanical preparation** through **abrasive blasting** or **grinding** creates **reactive surfaces**and**optimal profiles** for **coating application**.

Adhesion testing employs standardized methods including **cross-cut adhesion** (ASTM D3359), **pull-off adhesion**(ASTM D4541), and **knife adhesion** (ASTM D6677). These tests evaluate **adhesive strength** under **controlled conditions** but may not fully predict **long-term performance** under **service conditions**. **Environmental aging** tests including **salt spray**, **humidity cycling**, and **thermal shock** provide better **performance prediction** for **durability assessment**.

Opacity in paint systems results from **light scattering** and **absorption** by **pigment particles** dispersed in the **coating matrix**. **Rayleigh scattering** occurs when particle sizes are smaller than the **wavelength of light**, while **Mie scattering**dominates for particles comparable to or larger than the **light wavelength**. **Optimal opacity** typically occurs when **pigment particles** approximate **half the wavelength** of visible light.

The **Kubelka-Munk theory** provides the mathematical framework for **opacity calculations** in coating systems:

$$K/S = (1-R)^2/2R \qquad (1.12)$$

where K represents the **absorption coefficient**, S is the **scattering coefficient**, and R is the **reflectance** of the coating. This relationship enables **prediction of hiding power** based on **pigment properties** and **concentration**.

Pigment volume concentration (PVC) significantly influences **opacity** and other **coating properties**. The **critical pigment volume concentration** (CPVC) represents the point where **binder** just fills the **voids** between **pigment particles**. Below CPVC, coatings exhibit **maximum opacity** and **gloss**, while above CPVC, properties deteriorate due to **insufficient binder** for complete **pigment wetting**.

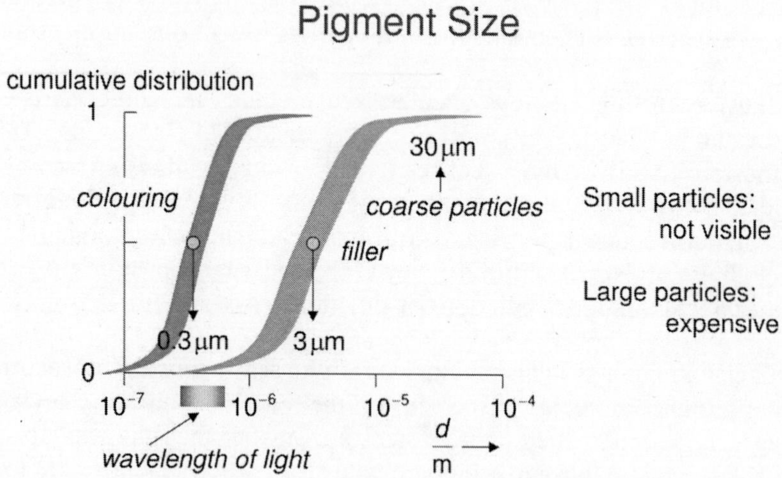

Fig. 5: Relationship between pigment volume concentration and coating properties showing critical pigment volume concentration effects

Fire Retardant and Luminescent Paints

Specialty coating systems incorporating **fire retardancy** and **luminescent properties** represent advanced applications of **chemical principles** in protective and functional coatings. These systems require **sophisticated formulation approaches**that balance **primary functionality** with **coating performance** and **environmental considerations.**

Fire retardant paints employ multiple **chemical mechanisms** to reduce **flammability** and **flame spread** of coated substrates. **Intumescent systems** represent the most common approach, utilizing **chemical reactions** that produce **insulating char layers** when exposed to heat. The **intumescent mechanism** involves **three key components**: an **acid source** (typically ammonium polyphosphate), a **carbon source** (pentaerythritol or similar polyol), and a **blowing agent**(melamine or dicyandiamide).

The **intumescent reaction sequence** begins with **thermal decomposition** of the **acid source**, generating **phosphoric acid** that catalyzes**dehydration** of the **carbon source**:

$$(NH_4PO_3)_n \rightarrow H_3PO_4 + NH_3 + H_2O \quad (1.13)$$

$$\text{Carbon source} + H_3PO_4 \rightarrow \text{Carbonaceous char} + H_2O \quad (1.14)$$

Simultaneously, the **blowing agent** decomposes to produce **non-combustible gases** that expand the **char structure**, creating an **insulating foam** with **low thermal conductivity**. This **expanded char layer** protects the underlying substrate by **reducing heat transfer** and **limiting oxygen access**.

Non-intumescent fire retardants include **halogenated compounds**, **metal hydroxides**, and **phosphorus-containing compounds** that interfere with **combustion reactions** through **radical scavenging** and **thermal quenching. Aluminum trihydrate** (ATH) and **magnesium hydroxide** function through **endothermic decomposition**, releasing **water vapor** that dilutes **combustible gases** and cools the **combustion zone**.

Luminescent paints incorporate **phosphorescent** or **fluorescent materials** that absorb **electromagnetic radiation** and re-emit **visible light**. **Photoluminescent systems** using **zinc sulfide** or **strontium aluminate** phosphors provide **long-lasting glow** after **light exposure**, while **radioluminescent systems** maintain **continuous emission** through **radioactive decay** of incorporated isotopes.

Strontium aluminate phosphors ($SrAl_2O_4:Eu^{2+},Dy^{3+}$) have largely replaced **zinc sulfide systems** due to their **superior brightness** and **extended afterglow duration**. The **europium ions** serve as **luminescent centers**, while **dysprosium** acts as a **co-activator** that creates **electron traps** responsible for **persistent phosphorescence**. The **crystal structure** and **dopant concentrations** determine **emission wavelength** and **decay characteristics**.

Encapsulation strategies protect **luminescent materials** from **moisture, chemical attack**, and **mechanical damage** that can reduce **performance** and **durability**. **Silica coatings, polymer encapsulation**, and **glass matrix systems** provide **barrier protection** while maintaining **optical transparency** necessary for **light transmission**.

Solvents, Thinners, and Their Toxicological Aspects

Solvent systems in paint formulations serve multiple **critical functions** including **viscosity control**, **application properties**, **film formation**, and **surface wetting**. The **selection** and **optimization** of solvent blends requires careful consideration of **evaporation rates**, **solvency parameters**, **environmental impact**, and **toxicological properties**. **Regulatory pressures** have driven significant **reformulation efforts** toward **lower-VOC** and **safer solvent systems**.

Solvent classification is based on **chemical structure** and **functional properties**. **Hydrocarbon solvents** including **aliphatic**, **aromatic**, and **naphthenic types** provide **good solvency** for many **resin systems** with **relatively low cost**. **Oxygenated solvents** such as **alcohols**, **ketones**, **esters**, and **ethers** offer **enhanced solvency** and **faster evaporation rates** but may present **increased health concerns**.

Hansen solubility parameters provide a **quantitative framework** for **solvent selection** based on **dispersion forces**(δD), **polar interactions** (δP), and **hydrogen bonding** (δH). The **total solubility parameter** is calculated as:

$$\delta_t^2 = \delta D^2 + \delta P^2 + \delta H^2 \tag{1.15}$$

Solvent blending utilizes these parameters to optimize **solvency**, **evaporation characteristics**, and **application properties**. **Three-dimensional solubility plots** enable **prediction** of **polymer-solvent compatibility** and **formulation optimization**.

Evaporation rate control involves blending **solvents** with different **vapor pressures** to achieve **desired drying characteristics**. **Fast-evaporating solvents** enable **rapid surface skinning** but may cause **surface defects**, while **slow solvents** improve **flow and leveling** but extend **drying times**. **Balanced solvent systems** combine **multiple componentsto** optimize both **application** and **final properties**.

Toxicological considerations encompass **acute toxicity**, **chronic health effects**, **reproductive toxicity**, and **environmental impact**. **Aromatic hydrocarbons** such as **benzene**, **toluene**, and **xylene** present **neurotoxic concernsand** potential **carcinogenicity**, leading to **regulatory restrictions** and **substitution efforts**. **Glycol ethers**

demonstrate **reproductive toxicity**, particularly **ethylene glycol derivatives**, driving adoption of **propylene glycol alternatives.**

Exposure assessment involves evaluation of **inhalation, dermal contact**, and **ingestion pathways** during **manufacturing, application**, and **service life. Vapor pressure, skin penetration rates**, and **bioaccumulation potential**influence**exposure levels** and **risk characterization. Personal protective equipment** (PPE) requirements and **engineering controls** are designed based on **toxicological data** and **exposure modeling.**

Regulatory frameworks including **REACH** (Registration, Evaluation, Authorization, and Restriction of Chemicals), **OSHA** (Occupational Safety and Health Administration), and **EPA** (Environmental Protection Agency) establish **exposure limits, labeling requirements**, and **use restrictions** for **solvent systems. VOC regulations** have driven development of **waterborne, high-solids**, and **powder coating systems** that reduce **organic solvent content.**

Green solvent alternatives include **bio-based solvents, ionic liquids**, and **supercritical fluids** that offer **reduced environmental impact** and **improved safety profiles. D-limonene, methyl soyate**, and **succinic acid derivatives**provide**renewable alternatives** to **petroleum-based solvents** while maintaining **acceptable performance** in many applications.

Basic Azo Dye Chemistry and Application on Textiles

Azo dyes constitute the largest class of **synthetic colorants**, representing approximately **70% of all commercial dyes**due to their **versatility, bright colors**, and **relatively simple synthesis**. The **fundamental chromophore** consists of the **azo group** (-N=N-) linking **aromatic ring systems**, creating **extended conjugation** that produces **intense visible absorption** and **color.**

The **general structure** of azo dyes can be represented as **Ar_1-N=N-Ar_2**, where Ar_1 and Ar_2 represent **aromatic systems**that may contain various **substituents** influencing **color, solubility**, and **application properties. Electron-donating groups** such as **amino** and **hydroxyl** groups typically produce **bathochromic shifts** (red-shift), while **electron-withdrawing groups** like **nitro** and **sulfonic acid** groups cause **hypsochromic shifts** (blue-shift).

Azo dye synthesis involves **diazotization** of **primary aromatic amines** followed by **coupling reactions** with **activated aromatic compounds**. The **diazotization reaction** occurs under **acidic conditions** at **low temperatures** to form **diazonium salts:**

$$ArNH_2 + HNO_2 + HCl \rightarrow ArN_2^+Cl^- + 2H_2O \quad (1.16)$$

The **coupling reaction** joins the **diazonium salt** with **coupling components** such as **phenols, naphthols**, or **aromatic amines** under **alkaline conditions**:

$$ArN_2^+ + Ar'H \rightarrow Ar\text{-}N{=}N\text{-}Ar' + H^+ \quad (1.17)$$

Textile application of azo dyes encompasses multiple **application classes** including **direct dyes, reactive dyes, acid dyes**, and **disperse dyes**, each designed for specific **fiber types** and **application methods. Direct dyes** demonstrate **substantivity** toward **cellulosic fibers** through **hydrogen bonding** and **van der Waals interactions**, enabling **simple application** from **aqueous baths** without **chemical modification**.

Reactive dyes contain **reactive groups** that form **covalent bonds** with **fiber molecules**, providing **superior wash fastness** and **color brilliance**. Common **reactive systems** include **vinyl sulfone, chlorotriazine,** and **monofluorotriazine**groups that react with **nucleophilic sites** on **cellulose fibers**:

$$Dye\text{-}SO_2CH{=}CH_2 + Cell\text{-}OH \rightarrow Dye\text{-}SO_2CH_2CH_2\text{-}O\text{-}Cell \quad (1.18)$$

Acid dyes utilize **ionic interactions** between **anionic dye molecules** and **cationic sites** on **protein fibers** such as **wool**and**silk**. The **pH** of the dyeing bath significantly influences **uptake rates** and **levelness**, with **optimal conditions**typically achieved at **pH 2-4** where **amino groups** in proteins are **protonated**.

Disperse dyes are specifically designed for **hydrophobic synthetic fibers** such as **polyester** and **acetate**. These **non-ionic dyes** exhibit **limited water solubility** and penetrate **fiber structures** through **dissolution** and **diffusion** processes at **elevated temperatures**. The **thermodynamic driving force** for **dye uptake** involves **reduction** of **interfacial energy**between**dye** and **fiber**.

Fastness properties including **light fastness**, **wash fastness**, and **crock fastness** depend on **molecular structure**, **aggregation behavior**, and **fiber-dye interactions**. **Planar molecular structures** tend to exhibit **better light fastness**through**efficient energy dissipation**, while **bulky substituents** may reduce **penetration** but improve **wash fastness**through**increased molecular size**.

Environmental considerations surrounding azo dye usage include **biodegradation pathways**, **toxicity assessment**, and **wastewater treatment** requirements. **Reductive cleavage** of **azo bonds** can produce **aromatic amines**, some of which demonstrate **carcinogenic properties**. **Advanced oxidation processes** using **ozone**, **hydrogen peroxide**, and **photocatalysis** provide effective **degradation methods** for **azo dye removal** from **industrial effluents**.

Table 1: Comparison of major azo dye classes for textile applications

Dye Class	Fiber Type	Application Method	Fastness Properties	Typical Structures
Direct	Cellulose	Aqueous bath	Moderate wash fastness	Sulfonated azo compounds
Reactive	Cellulose	Alkaline conditions	Excellent wash fastness	Reactive group-containing azo dyes
Acid	Protein	Acidic bath	Good light fastness	Sulfonated structures
Disperse	Synthetic	High temperature	Excellent sublimation fastness	Non-ionic azo compounds

Conclusion

This chapter has provided a comprehensive examination of **chemical applications** in food analysis and industrial systems, demonstrating the **interconnected nature** of analytical chemistry across diverse sectors. The exploration of **dairy chemistry** reveals the complex **colloidal**

systems that require sophisticated analytical approaches to ensure **quality**and**safety**. **Modern instrumental methods** including **HPLC**, **GC-MS**, and **ICP-MS** have revolutionized our ability to characterize **food components** and **contaminants** at unprecedented **sensitivity levels**.

The analysis of **beverages** and **food additives** illustrates the critical role of **chemical analysis** in **regulatory compliance**and**consumer protection**. **Caffeine quantification**, **adulterant detection**, and **preservative analysis** demonstrate how **analytical chemistry** supports both **quality assurance** and **safety assessment** in complex food matrices. The **integration**of classical methods with modern instrumental techniques provides**robust analytical frameworks** for **routine monitoring** and **research applications**.

Colorant analysis and **pesticide residue determination** highlight the **global challenges** facing food safety laboratories in addressing **contamination** from both **intentional additives** and **environmental sources**. The **multi-residue capabilities** of **GC-MS** and **LC-MS/MS** enable **comprehensive screening** programs that support **international trade**and**public health protection**.

The examination of **industrial applications** including **paints**, **pigments**, and **dyes** reveals the **broader context** of chemical principles that extend beyond food systems. **Understanding** these connections enhances appreciation for the **fundamental chemistry** underlying diverse industrial processes and provides insight into **environmental fate** and **exposure pathways** for chemical substances.

Future developments in this field will likely emphasize **green analytical chemistry**, **rapid screening methods**, and **portable instrumentation** that enable **field-based analysis**. The **integration** of **artificial intelligence** and **machine learning** with **analytical data** promises to enhance **method development**, **data interpretation**, and **quality control**processes. **Continued advancement** in **analytical capabilities** will support **emerging challenges** in **food security**, **authenticity verification**, and **environmental monitoring**, ensuring that **chemical analysis** remains central to **protecting public health** and **maintaining food quality** in an increasingly complex global marketplace.

Key Takeaways:

- **Modern analytical chemistry** provides the essential tools for **food safety assessment** and **quality control** across diverse food systems
- **Multi-residue methods** using **advanced instrumentation** enable **comprehensive monitoring** of **contaminants** and **additives** in complex matrices
- **Integration** of **classical** and **instrumental methods** provides **robust analytical frameworks** for **routine** and **research applications**
- **Industrial chemical applications** demonstrate the **broader relevance** of analytical principles beyond food systems
- **Future developments** will emphasize **green chemistry**, **rapid methods**, and **intelligent data analysis** to address **emerging challenges** in food safety and environmental protection.

CHAPTER 2
ENVIRONMENTAL CHEMISTRY AND POLLUTION CONTROL

Learning Objectives

Upon completion of this chapter, students will be able to:

- Analyze the chemical mechanisms underlying air pollution formation and understand the role of catalytic converters in emission control
- Evaluate water quality parameters using analytical techniques and interpret their significance for public health
- Assess the effectiveness of wastewater treatment processes and their environmental impact
- Examine soil chemistry fundamentals and the role of fertilizers in nutrient cycling • Apply chemical principles to design sustainable pollution control strategies

Introduction

Environmental chemistry represents one of the most critical applications of chemical principles in addressing global challenges of the 21st century. This interdisciplinary field combines fundamental chemical knowledge with environmental science to understand, monitor, and mitigate the impact of human activities on air, water, and soil systems. The increasing industrialization and urbanization have led to unprecedented levels of pollution, making environmental chemistry essential for developing sustainable solutions.

The chemical basis of environmental processes involves complex interactions between natural and anthropogenic compounds across multiple phases and scales. **Atmospheric chemistry** governs the formation of greenhouse gases, acid rain, and photochemical smog, while **aquatic chemistry** determines water quality and the fate of pollutants in hydrological systems. Similarly, **soil chemistry** influences nutrient availability, contaminant mobility, and ecosystem health.

Modern environmental chemistry emerged in the 1960s following Rachel Carson's seminal work "Silent Spring" (1962), which highlighted the environmental consequences of chemical pesticides.

Since then, the field has evolved to encompass advanced analytical techniques, mechanistic understanding of pollution processes, and innovative treatment technologies. The development of environmental regulations such as the Clean Air Act (1970) and the Safe Drinking Water Act (1974) in the United States, along with similar legislation worldwide, has driven the need for rigorous chemical analysis and monitoring.

Contemporary environmental challenges require a comprehensive understanding of chemical processes at molecular, local, and global scales. **Climate change** involves complex atmospheric chemistry of greenhouse gases, while **water scarcity** demands advanced treatment technologies based on chemical and physical processes. The concept of **green chemistry**, introduced by Paul Anastas and John Warner in the 1990s, emphasizes the design of chemical products and processes that reduce or eliminate hazardous substances.

This chapter provides a systematic exploration of environmental chemistry principles, beginning with atmospheric processes and air pollution control, progressing through water chemistry and quality assessment, and concluding with soil chemistry and sustainable agricultural practices. Each section emphasizes both theoretical understanding and practical applications, preparing students to address real-world environmental challenges through chemical science.

2.1 Air Pollution Chemistry and Mitigation

Classification of Air Pollutants

Air pollution represents a complex mixture of chemical species that originate from both natural and anthropogenic sources. The **World Health Organization (WHO)** defines air pollution as contamination of the indoor or outdoor environment by any chemical, physical, or biological agent that modifies the natural characteristics of the atmosphere. Understanding the chemical nature of air pollutants is fundamental to developing effective control strategies and assessing their environmental and health impacts.

Primary pollutants are emitted directly into the atmosphere from identifiable sources, while **secondary pollutants** form through chemical reactions involving primary pollutants and atmospheric constituents. This distinction is crucial for pollution control strategies, as reducing primary emissions may not immediately translate to proportional reductions in secondary pollutants due to complex atmospheric chemistry.

Particulate matter (PM) represents one of the most significant air quality concerns globally. PM_{10} (particles with aerodynamic diameter ≤ 10 μm) and $PM_{2.5}$ (particles ≤ 2.5 μm) differ substantially in their chemical composition, sources, and health effects. Coarse particles ($PM_{10-2.5}$) primarily originate from mechanical processes such as wind erosion, road dust, and construction activities, consisting mainly of crustal materials like silicates, carbonates, and metal oxides. Fine particles ($PM_{2.5}$) result predominantly from combustion processes and atmospheric chemical reactions, containing significant fractions of organic carbon, elemental carbon, sulfates, nitrates, and ammonium compounds.

The chemical composition of $PM_{2.5}$ varies significantly by location and season, reflecting differences in emission sources and atmospheric processing. Urban areas typically show higher concentrations of elemental carbon from vehicle emissions and organic compounds from fossil fuel combustion, while rural areas may exhibit elevated concentrations of secondary organic aerosols formed from biogenic volatile organic compounds (VOCs) emitted by vegetation.

Gaseous pollutants encompass a diverse range of chemical species with varying atmospheric lifetimes and environmental impacts. **Sulfur dioxide (SO_2)** primarily originates from fossil fuel combustion, particularly coal-fired power plants and industrial processes. The atmospheric oxidation of SO_2 leads to sulfuric acid formation through both homogeneous and heterogeneous pathways:

$$SO_2 + OH\bullet \rightarrow HSO_3\bullet \quad (2.1)$$

$$HSO_3\bullet + O_2 \rightarrow SO_3 + HO_2\bullet \quad (2.2)$$

$$SO_3 + H_2O \rightarrow H_2SO_4 \quad (2.3)$$

Nitrogen oxides ($NO_x = NO + NO_2$) play central roles in atmospheric chemistry as precursors to ozone formation and contributors to acid deposition. The primary emission source is high-temperature combustion processes, where atmospheric nitrogen and oxygen react according to the Zeldovich mechanism. The subsequent atmospheric chemistry involves rapid interconversion between NO and NO_2 through photochemical processes, making NO_x a key component in urban air quality management.

Volatile organic compounds (VOCs) represent a diverse class of carbon-containing chemicals that significantly influence atmospheric chemistry. Biogenic VOCs, including isoprene (C_5H_8) and monoterpenes ($C_{10}H_{16}$), constitute the largest global VOC emissions, exceeding anthropogenic sources by approximately a factor of ten. However, anthropogenic VOCs, particularly in urban areas, play disproportionate roles in ozone formation due to their higher reactivity and co-location with NO_x emissions.

The **reactivity** of different VOCs varies dramatically, influencing their relative importance in ozone formation. The concept of **photochemical ozone creation potential (POCP)** provides a quantitative framework for comparing the ozone-forming potential of different VOCs. Highly reactive compounds such as alkenes and aromatics have much higher POCP values than alkanes, leading to differentiated regulatory approaches for different VOC classes.

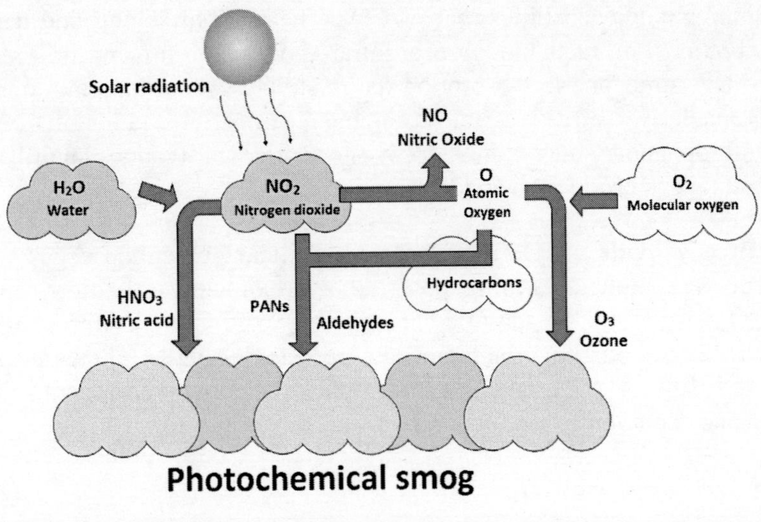

Fig. 6: Atmospheric transformation pathways of major air pollutants showing primary emissions, secondary formation processes, and removal mechanisms

Greenhouse Gases and Acid Rain Formation

The enhanced greenhouse effect represents one of the most pressing environmental challenges of our time, driven by increasing atmospheric concentrations of greenhouse gases from human activities. The radiative forcing potential of different greenhouse gases depends on their **absorption spectra** in the infrared region and their **atmospheric lifetimes**. Understanding the molecular basis of radiative forcing provides essential insights for climate mitigation strategies.

Carbon dioxide (CO_2) constitutes the most significant anthropogenic greenhouse gas, accounting for approximately 76% of total greenhouse gas emissions. The atmospheric concentration of CO_2 has increased from pre-industrial levels of 280 ppm to over 410 ppm as of 2023, primarily due to fossil fuel combustion and land-use changes. The molecular structure of CO_2 allows absorption of infrared radiation at wavelengths around 15 μm and 4.3 μm, corresponding to bending and stretching vibrational modes, respectively.

Methane (CH_4) exhibits a much higher global warming potential than CO_2 on a per-molecule basis, with a 100-year global warming potential approximately 25 times that of CO_2. The C-H stretching and bending vibrations in methane absorb infrared radiation in spectral regions where atmospheric absorption is relatively weak, making methane particularly effective at trapping terrestrial radiation. Major anthropogenic sources include livestock, rice cultivation, landfills, and natural gas production and distribution.

Nitrous oxide (N_2O) represents a significant greenhouse gas with a 100-year global warming potential approximately 298 times that of CO_2. Agricultural activities, particularly the use of nitrogen-based fertilizers, constitute the largest anthropogenic source. The atmospheric chemistry of N_2O involves stratospheric reactions that contribute to ozone depletion:

$$N_2O + O(^1D) \rightarrow 2NO \quad (2.4)$$

$$NO + O_3 \rightarrow NO_2 + O_2 \quad (2.5)$$

Acid rain formation results from the atmospheric oxidation of sulfur and nitrogen compounds to form strong acids that subsequently deposit to terrestrial and aquatic ecosystems. The term "acid rain" encompasses all forms of acidic deposition, including wet deposition (rain, snow, fog) and dry deposition (gases and particles). The pH of unpolluted rainwater is approximately 5.6 due to equilibrium with atmospheric CO_2, but acid rain can reach pH values below 4.0 in severely polluted regions.

The aqueous-phase oxidation of SO_2 in cloud droplets represents a major pathway for sulfuric acid formation, particularly in the presence of catalysts such as iron and manganese ions. The rate of SO_2 oxidation increases dramatically with decreasing pH, creating a positive feedback mechanism that accelerates acid formation in already acidic droplets. This process can be represented by the overall reaction:

$$SO_2 + H_2O_2 \rightarrow H_2SO_4 \quad (2.6)$$

Nitric acid (HNO_3) formation occurs primarily through gas-phase reactions involving NO_2 and hydroxyl radicals. The daytime formation pathway involves:

$$NO_2 + OH\bullet \rightarrow HNO_3 \quad (2.7)$$

Nighttime formation occurs through the hydrolysis of dinitrogen pentoxide (N_2O_5), which forms from NO_2 and NO_3 radicals:

$$NO_2 + NO_3\bullet \rightarrow N_2O_5 \quad (2.8)$$

$$N_2O_5 + H_2O \rightarrow 2HNO_3 \quad (2.9)$$

The environmental impacts of acid deposition include **soil acidification**, leading to aluminum mobilization and base cation leaching; **forest decline** through direct foliar damage and nutrient imbalances; and **aquatic ecosystem acidification**, resulting in fish kills and biodiversity loss. The **critical loads** concept, developed in Europe during the 1980s, provides a quantitative framework for assessing ecosystem sensitivity to acid deposition and establishing emission reduction targets.

CFCs and Ozone Depletion Mechanisms

The discovery of **stratospheric ozone depletion** by chlorofluorocarbons (CFCs) represents one of the most significant environmental chemistry discoveries of the 20th century. The work of

Mario Molina and F. Sherwood Rowland in 1974 demonstrated that CFCs, widely used as refrigerants and aerosol propellants, could catalytically destroy stratospheric ozone through a series of chemical reactions. This discovery led to the Montreal Protocol (1987), considered one of the most successful international environmental agreements.

Chlorofluorocarbons are synthetic compounds consisting of carbon, chlorine, and fluorine atoms. Common CFCs include CFC-11 (CCl_3F), CFC-12 (CCl_2F_2), and CFC-113 (CCl_2FCClF_2). These compounds are chemically inert in the troposphere due to the absence of natural removal processes, allowing them to survive for decades to centuries before reaching the stratosphere through vertical mixing.

The **stratospheric ozone layer**, located primarily between 15 and 35 km altitude, absorbs ultraviolet radiation in the UV-B (280-315 nm) and UV-C (200-280 nm) ranges, protecting life on Earth from harmful solar radiation. The natural ozone chemistry involves the **Chapman cycle**, where ozone formation and destruction occur through oxygen-only reactions:

$$O_2 + h\nu \rightarrow 2O \tag{2.10}$$

$$O + O_2 + M \rightarrow O_3 + M \tag{2.11}$$

$$O_3 + h\nu \rightarrow O_2 + O \tag{2.12}$$

$$O + O_3 \rightarrow 2O_2 \tag{2.13}$$

However, this simplified mechanism overestimates stratospheric ozone concentrations, leading to the recognition that **catalytic cycles** involving trace species significantly affect ozone abundances. The general form of catalytic ozone destruction involves a catalyst X that is regenerated:

$$X + O_3 \rightarrow XO + O_2 \tag{2.14}$$

$$XO + O \rightarrow X + O_2 \tag{2.15}$$

$$\text{Net: } O_3 + O \rightarrow 2O_2 \tag{2.16}$$

CFC photolysis in the stratosphere releases chlorine atoms that initiate catalytic ozone destruction cycles. The UV radiation at wavelengths below 220 nm photolyzes CFCs:

$$CCl_3F + h\nu \rightarrow CCl_2F + Cl \tag{2.17}$$

The released chlorine atoms then participate in catalytic ozone destruction:

$$Cl + O_3 \rightarrow ClO + O_2 \tag{2.18}$$

$$ClO + O \rightarrow Cl + O_2 \tag{2.19}$$

$$\text{Net: } O_3 + O \rightarrow 2O_2 \tag{2.20}$$

A single chlorine atom can destroy thousands of ozone molecules before being removed from the catalytic cycle through reactions that form **reservoir species** such as hydrogen chloride (HCl) and chlorine nitrate ($ClONO_2$).

The **Antarctic ozone hole**, discovered by British Antarctic Survey scientists in 1985, revealed the existence of additional ozone depletion mechanisms involving **polar stratospheric clouds (PSCs)**. During the Antarctic winter, extremely low temperatures (below -78°C) allow the formation of PSCs composed of nitric acid trihydrate and water ice. **Heterogeneous reactions** on PSC surfaces convert reservoir species to more reactive forms:

$$ClONO_2 + HCl \rightarrow Cl_2 + HNO_3 \tag{2.21}$$

When sunlight returns in the Antarctic spring, the Cl_2 photolyzes rapidly, releasing chlorine atoms that initiate rapid ozone destruction through the **ClO-ClO cycle**:

$$Cl_2 + h\nu \rightarrow 2Cl \tag{2.22}$$

$$2(Cl + O_3 \rightarrow ClO + O_2) \tag{2.23}$$

$$ClO + ClO + M \rightarrow Cl_2O_2 + M \tag{2.24}$$

$$Cl_2O_2 + h\nu \rightarrow Cl + ClOO \tag{2.25}$$

$$ClOO + M \rightarrow Cl + O_2 + M \tag{2.26}$$

Net: $2O_3 \rightarrow 3O_2$ (2.27)

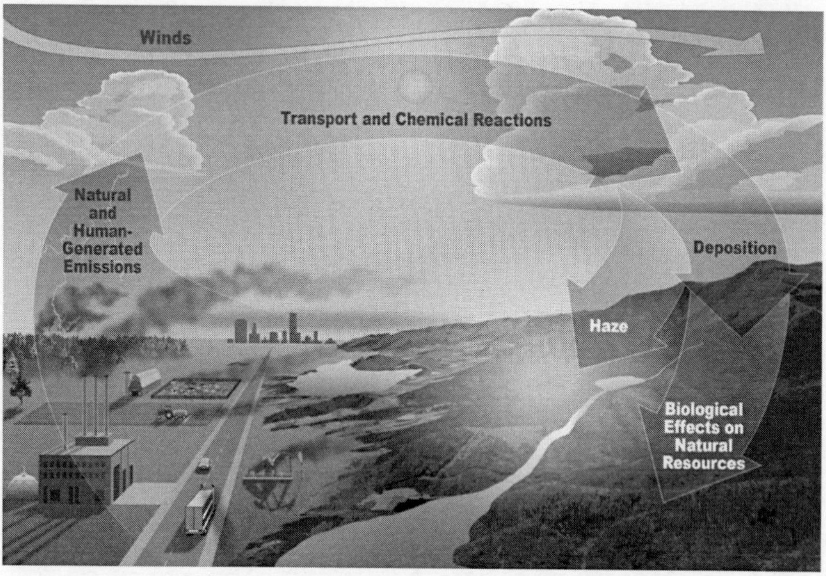

Fig. 7: Schematic diagram of stratospheric ozone chemistry showing natural formation/destruction cycles and anthropogenic catalytic depletion mechanisms

PAN and Photochemical Smog Dynamics

Photochemical smog represents a complex secondary air pollution phenomenon characterized by elevated concentrations of ozone and other oxidants formed through photochemical reactions involving volatile organic compounds (VOCs) and nitrogen oxides (NO_x) in the presence of sunlight. This type of air pollution was first recognized in Los Angeles during the 1940s and has since become a widespread urban air quality problem worldwide.

Peroxyacetyl nitrate (PAN) serves as both an important component of photochemical smog and a key indicator of photochemical activity in the atmosphere. PAN formation occurs through the reaction of peroxyacetyl radicals with nitrogen dioxide:

$$CH_3CO\text{-}O_2\bullet + NO_2 \leftrightarrow CH_3CO\text{-}O_2NO_2 \quad (2.28)$$

The formation of peroxyacetyl radicals involves the oxidation of acetaldehyde by hydroxyl radicals, followed by reaction with molecular oxygen:

$$CH_3CHO + OH\bullet \rightarrow CH_3CO\bullet + H_2O \quad (2.29)$$

$$CH_3CO\bullet + O_2 \rightarrow CH_3CO\text{-}O_2\bullet \quad (2.30)$$

PAN exhibits **thermal instability** at temperatures above 15°C, with a lifetime of approximately 30 minutes at 25°C. This temperature dependence allows PAN to serve as a **reservoir species** for NO_x, transporting nitrogen oxides from urban source regions to remote areas where thermal decomposition releases NO_2 and contributes to rural ozone formation.

The **fundamental mechanism** of photochemical smog formation involves complex interactions between VOCs, NO_x, and sunlight. The process begins with the photolysis of nitrogen dioxide, the only significant photolytic source of atomic oxygen in the lower atmosphere:

$$NO_2 + h\nu\ (\lambda < 420\ \text{nm}) \rightarrow NO + O \quad (2.31)$$

The atomic oxygen rapidly reacts with molecular oxygen to form ozone:

$$O + O_2 + M \rightarrow O_3 + M \quad (2.32)$$

In the absence of VOCs, ozone formation is limited by the rapid reaction between NO and O_3:

$$NO + O_3 \rightarrow NO_2 + O_2 \quad (2.33)$$

This reaction establishes a **photostationary state** where ozone concentrations remain relatively low. However, the presence of VOCs fundamentally alters this chemistry by providing alternative pathways for NO oxidation to NO_2 that do not consume ozone.

VOC oxidation is initiated primarily by reaction with hydroxyl radicals (OH•), the most important oxidant in the troposphere. The hydroxyl radical concentration shows a strong diurnal cycle, with maximum values occurring around midday when photochemical activity is highest. VOC oxidation produces **peroxy radicals ($RO_2\bullet$)** that can oxidize NO to NO_2 without consuming ozone:

$$RO_2\bullet + NO \rightarrow RO\bullet + NO_2 \quad (2.34)$$

This reaction provides the key mechanism for **net ozone formation** in photochemical smog episodes. The organic peroxy radicals formed from VOC oxidation are numerous and diverse, reflecting the chemical complexity of VOC emissions from mobile sources, industrial activities, and biogenic emissions.

The **ozone formation potential** of different VOCs varies significantly based on their reaction rates with OH• radicals and the yields of peroxy radicals from their oxidation. **Aromatic hydrocarbons** such as toluene and xylene exhibit particularly high ozone formation potentials due to their rapid reaction rates and complex oxidation mechanisms that produce multiple peroxy radicals per molecule oxidized.

Aldehyde formation represents an important consequence of VOC oxidation in photochemical smog. Formaldehyde (HCHO) and acetaldehyde (CH_3CHO) are primary products of many VOC oxidation pathways and contribute to both PAN formation and continued radical production through photolysis:

$$HCHO + h\nu \rightarrow HCO\bullet + H\bullet \quad (2.35)$$

$$HCO\bullet + O_2 \rightarrow HO_2\bullet + CO \quad (2.36)$$

The **NO_x-VOC sensitivity** of ozone formation represents a critical consideration for air quality management strategies. In VOC-limited conditions, typical of many urban cores, ozone formation is most effectively controlled by reducing VOC emissions. In NO_x-limited conditions, more characteristic of rural and suburban areas, NO_x reductions provide the most effective ozone control strategy.

Regional transport of photochemical smog precursors and products complicates local air quality management efforts. Ozone and its precursors can travel hundreds of kilometers from source regions, contributing to elevated ozone concentrations in downwind rural and suburban areas. This regional character of photochemical smog has led to coordinated multi-state and international control strategies.

Mobile Source Pollution and Catalytic Converters

Mobile sources, including light-duty vehicles, heavy-duty trucks, aircraft, ships, and off-road equipment, represent major contributors to urban air pollution worldwide. The composition of mobile source emissions varies significantly among different vehicle types, fuel compositions, and operating conditions. Understanding the chemical basis of mobile source emissions is essential for developing effective emission control technologies and fuel formulation strategies.

Internal combustion engines produce pollutant emissions through several mechanisms. **Incomplete combustion** results from insufficient oxygen availability, inadequate mixing, or insufficient residence time at high temperatures. This leads to emissions of carbon monoxide (CO), unburned hydrocarbons (HC), and particulate matter. **High-temperature combustion** promotes thermal NO_x formation through the Zeldovich mechanism, where atmospheric nitrogen and oxygen react at temperatures above 1300°C.

The **three-way catalytic converter** represents one of the most successful pollution control technologies, capable of simultaneously reducing emissions of CO, HC, and NO_x from gasoline-fueled vehicles. The term "three-way" refers to the catalyst's ability to promote three distinct chemical reactions:

$$2CO + O_2 \rightarrow 2CO_2 \qquad (2.37)$$

$$C_xH_\gamma + (x + y/4)O_2 \rightarrow xCO_2 + (y/2)H_2O \qquad (2.38)$$

$$2NO + 2CO \rightarrow N_2 + 2CO_2 \qquad (2.39)$$

The **catalyst composition** typically consists of platinum, palladium, and rhodium supported on a high-surface-area washcoat of aluminum oxide (Al_2O_3) applied to a ceramic or metallic monolith substrate. **Platinum and palladium**promote oxidation reactions of CO and HC, while **rhodium** is particularly effective for NO_x reduction reactions. The noble metals are dispersed at the nanometer scale to maximize their surface area and catalytic activity.

Oxygen storage materials, primarily cerium oxide (CeO_2) and cerium-zirconium mixed oxides, play crucial roles in three-way catalyst performance. These materials can reversibly store and release oxygen based on the local oxygen concentration:

$$2CeO_2 \leftrightarrow Ce_2O_3 + \frac{1}{2}O_2 \quad (2.40)$$

This oxygen storage capacity allows the catalyst to function effectively under the slight variations in air-fuel ratio that occur during real-world driving conditions. During fuel-rich conditions, stored oxygen is released to promote CO and HC oxidation, while during fuel-lean conditions, excess oxygen is stored to maintain NO_x reduction activity.

The **light-off temperature** represents a critical parameter for catalytic converter performance, defined as the temperature at which 50% conversion efficiency is achieved. Modern three-way catalysts typically achieve light-off temperatures around 200-300°C for CO and HC oxidation and 250-350°C for NO_x reduction. **Cold-start emissions** during the first few minutes of vehicle operation, before the catalyst reaches its operating temperature, contribute disproportionately to total vehicle emissions.

Catalyst deactivation mechanisms include thermal degradation, poisoning, and fouling. **Thermal degradation** occurs through sintering of the noble metal particles and loss of washcoat surface area at high exhaust temperatures. **Poisoning**involves the irreversible adsorption of contaminants such as sulfur, phosphorus, and lead that block active catalytic sites. **Fouling** results from the physical deposition of particles and oil-derived materials on the catalyst surface.

Diesel engine emissions present different challenges compared to gasoline engines due to the fundamentally different combustion process. Diesel engines operate under fuel-lean conditions with excess air, producing lower concentrations of CO and HC but higher concentrations of NO_x and particulate matter. The oxygen-rich exhaust environment precludes the use of three-way catalysts for NO_x control.

Selective catalytic reduction (SCR) technology addresses NO_x emissions from diesel engines through the injection of a reductant, typically urea ($(NH_2)_2CO$), into the exhaust stream. The urea undergoes thermal decomposition and hydrolysis to produce ammonia:

$$(NH_2)_2CO \rightarrow NH_3 + HNCO \quad (2.41)$$

$$HNCO + H_2O \rightarrow NH_3 + CO_2 \quad (2.42)$$

The ammonia then reacts with NO_x over a catalyst, typically vanadium oxide supported on titanium dioxide (V_2O_5/TiO_2), to produce nitrogen and water:

$$4NH_3 + 4NO + O_2 \rightarrow 4N_2 + 6H_2O \quad (2.43)$$

$$8NH_3 + 6NO_2 \rightarrow 7N_2 + 12H_2O \quad (2.44)$$

Diesel particulate filters (DPF) address particulate matter emissions through physical filtration and periodic regeneration. The filter substrate, typically silicon carbide or cordierite, traps particulate matter while allowing exhaust gases to pass through porous walls. **Regeneration** involves the oxidation of trapped soot particles at temperatures above 550°C:

$$C + O_2 \rightarrow CO_2 \quad (2.45)$$

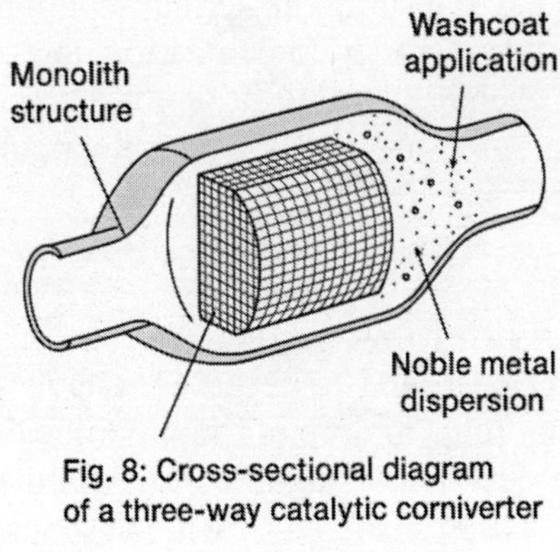

Fig. 8: Cross-sectional diagram of a three-way catalytic converter showing monolith structure, washcoat application, and noble metal dispersion

2.2 Water Chemistry and Quality Assessment

Hydrologic Cycle Overview

The **hydrologic cycle** represents the continuous movement of water through Earth's atmosphere, land surface, and subsurface, driven by solar energy and gravitational forces. This global process governs the distribution and quality of freshwater resources, influencing both natural ecosystem functioning and human water supply systems. Understanding the chemical transformations that occur during each phase of the hydrologic cycle is fundamental to water quality assessment and management.

Evaporation from oceans, lakes, rivers, and soil surfaces represents the primary mechanism by which water enters the atmosphere. This phase change requires approximately 2.45 kJ/g of energy at 20°C, making evaporation a major component of Earth's energy balance. The evaporation process naturally **purifies water** by leaving dissolved salts and other non-volatile substances behind, although some volatile organic compounds and certain metals can transfer to the vapor phase under specific conditions.

Transpiration by vegetation contributes significantly to atmospheric water vapor, with plants extracting water from soil through their root systems and releasing it through stomatal pores in leaves. The combined process of evaporation and transpiration, termed **evapotranspiration**, accounts for approximately 60% of precipitation that falls on land surfaces globally. The chemical composition of transpired water is generally purer than the soil water from which it originates, as plant root systems and xylem transport mechanisms exclude most dissolved ions.

Atmospheric water chemistry involves complex interactions between water vapor, aerosol particles, and trace gases. **Cloud condensation nuclei (CCN)** play crucial roles in droplet formation, with the chemical composition of these particles significantly influencing cloud microphysics and precipitation chemistry. Sea salt aerosols contribute chloride and sodium ions to precipitation, particularly in coastal regions, while continental aerosols contribute sulfates, nitrates, and various trace elements.

The **residence time** of water in different components of the hydrologic cycle varies dramatically, from days to weeks in the atmosphere to thousands of years in deep groundwater aquifers. This variation in residence time has profound implications for water quality, as longer residence times allow greater opportunities for chemical reactions and contamination to occur. Surface water systems typically have residence times of days to years, making them more susceptible to short-term pollution events but also more responsive to remediation efforts.

Precipitation chemistry reflects the integration of atmospheric processes occurring over large spatial scales. **Rainwater composition** varies geographically and seasonally, influenced by local and regional emission sources, atmospheric transport patterns, and meteorological conditions. In remote areas far from anthropogenic sources, rainwater pH typically ranges from 5.0 to 5.6 due to equilibrium with atmospheric CO_2. However, in polluted regions, precipitation can become significantly more acidic due to the presence of sulfuric and nitric acids derived from air pollutant emissions.

Surface water chemistry evolves through interactions with the atmosphere, soil, rock, and biological systems. **Weathering reactions** involving carbonic acid (H_2CO_3) formed from dissolved CO_2 represent primary mechanisms for mineral dissolution and major ion generation:

$$CO_2 + H_2O \leftrightarrow H_2CO_3 \tag{2.46}$$

$$CaCO_3 + H_2CO_3 \leftrightarrow Ca^{2+} + 2HCO_3^- \tag{2.47}$$

These reactions establish the **alkalinity** and **hardness** characteristics of natural waters, with alkalinity representing the acid-neutralizing capacity and hardness reflecting the concentration of divalent cations, primarily calcium and magnesium.

Groundwater chemistry develops through prolonged contact with soil and rock materials during infiltration and subsurface flow. The chemical evolution of groundwater typically follows predictable patterns based on the geochemical environment and residence time. **Shallow groundwater** often exhibits relatively low dissolved solids concentrations and oxic conditions, while **deep groundwater** may become increasingly mineralized and develop reducing conditions that mobilize metals such as iron, manganese, and arsenic.

The **vadose zone**, representing the unsaturated region between the land surface and the water table, plays a critical role in groundwater quality protection. **Soil filtration** processes remove particles and many dissolved contaminants, while **biodegradation** by soil microorganisms can transform or mineralize organic pollutants. However, the vadose zone also provides pathways for contaminant migration to groundwater, particularly for persistent chemicals that resist biodegradation.

Water quality degradation can occur at any point in the hydrologic cycle through natural processes or anthropogenic activities. **Point sources** such as industrial discharge pipes and sewage treatment plants introduce localized contamination, while **non-point sources** including agricultural runoff and urban stormwater contribute diffuse pollution over large areas. The temporal and spatial scales of water quality impacts depend on the specific contaminants involved, the receiving water characteristics, and the hydrodynamic conditions.

Standards for Drinking Water (WHO/IS Standards)

Drinking water quality standards represent scientifically-based guidelines and regulations designed to protect public health from waterborne diseases and chemical contamination. These standards are developed by national and international organizations through comprehensive risk assessment processes that consider toxicological data, occurrence studies, treatment feasibility, and analytical capabilities. The **World Health Organization (WHO)** provides global guidance through its Guidelines for Drinking-water Quality, while individual countries establish legally enforceable standards adapted to local conditions and resources.

The **WHO Guidelines for Drinking-water Quality** establish health-based targets for chemical, microbiological, and radiological parameters in drinking water. These guidelines undergo periodic revision based on emerging scientific evidence, with the most recent edition (fourth edition, 2017) incorporating advances in toxicology, epidemiology, and risk assessment methodologies. The WHO approach emphasizes **risk-based management** rather than rigid numerical standards, recognizing that local conditions may require adaptation of the guidelines.

Microbiological quality represents the highest priority for drinking water safety, as waterborne infectious diseases pose immediate and severe health risks. The WHO guidelines specify that drinking water should not contain any **pathogenic microorganisms**, including bacteria, viruses, and parasites. However, direct detection of all potential pathogens is technically challenging and economically impractical, leading to the use of **indicator organisms** such as *E. coli* and coliform bacteria.

Escherichia coli serves as the preferred indicator for fecal contamination because it is exclusively found in the intestines of warm-blooded animals and demonstrates similar survival characteristics to many pathogenic bacteria. The WHO guideline value for *E. coli* is zero colony-forming units (CFU) per 100 mL of water. **Total coliform bacteria** provide a broader indicator of microbiological quality, reflecting both fecal contamination and natural occurrence in the environment.

Chemical quality standards address both **health-based parameters** that may cause adverse health effects and **aesthetic parameters** that affect taste, odor, or appearance but do not pose direct health risks. Health-based parameters include inorganic chemicals such as heavy metals and metalloids, organic chemicals including pesticides and industrial solvents, and disinfection by-products formed during water treatment processes.

Heavy metal contamination represents a significant concern for drinking water safety due to the persistence and bioaccumulation potential of these substances. **Lead** contamination typically occurs through dissolution from plumbing materials, particularly in older distribution systems with lead service lines or lead-soldered copper pipes. The WHO guideline value for lead is 10 μg/L, reflecting evidence of neurotoxic effects at low exposure levels, particularly in children.

Arsenic occurs naturally in groundwater in many regions worldwide through dissolution from arsenic-bearing minerals and rocks. Long-term exposure to arsenic through drinking water causes skin lesions, cardiovascular disease, and various cancers. The WHO guideline value of 10 μg/L represents a balance between health protection and analytical and treatment feasibility, as many water sources naturally exceed this level.

Nitrate contamination of drinking water sources primarily results from agricultural fertilizer application and sewage discharge. High nitrate concentrations pose particular risks to infants through the development of **methemoglobinemia**(blue baby syndrome), where nitrate is reduced to nitrite in the infant's digestive system and oxidizes hemoglobin to methemoglobin. The WHO guideline value for nitrate is 50 mg/L as NO_3^- (equivalent to 11.3 mg/L as N).

Fluoride occurs naturally in many water sources and is deliberately added to drinking water in some countries for dental caries prevention. However, excessive fluoride exposure causes **dental fluorosis** and, at higher levels, **skeletal fluorosis.** The WHO guideline value for fluoride is 1.5 mg/L, though optimal levels for dental health are typically 0.5-1.0 mg/L depending on climate and water consumption patterns.

Organic chemical contaminants in drinking water include a diverse range of synthetic chemicals such as pesticides, industrial solvents, and petroleum products. Many organic chemicals exhibit **endocrine disrupting properties** or **carcinogenic potential**, requiring extremely low guideline values. **Benzene**, a known human carcinogen commonly found in gasoline, has a WHO guideline value of 10 μg/L based on lifetime cancer risk calculations.

Pesticide residues in drinking water reflect widespread agricultural use and environmental persistence of these compounds. The WHO guidelines include specific values for numerous individual pesticides, typically ranging from 0.1 to 100 μg/L depending on toxicological properties. **Atrazine**, one of the most widely used herbicides globally, has a guideline value of 100 μg/L based on mammalian toxicity studies, though concerns about endocrine disruption continue to drive research and regulatory evaluation.

Disinfection by-products (DBPs) form during drinking water treatment through reactions between disinfectants and natural organic matter. **Trihalomethanes (THMs)** represent the most abundant class of DBPs, forming through reactions between chlorine and humic substances:

$$\text{Humic matter} + Cl_2 \rightarrow CHCl_3 + CHBr_2Cl + CHBrCl_2 + CHBr_3 \quad (2.48)$$

The WHO guideline for total trihalomethanes is 100 μg/L, based on evidence of liver toxicity and cancer risk from long-term exposure. **Haloacetic acids (HAAs)** represent another significant DBP class, with guideline values ranging from 20 to 200 μg/L for different compounds.

Indian Standards (IS) for drinking water, established by the Bureau of Indian Standards (BIS), provide nationally specific requirements that consider local water quality conditions and public health priorities. **IS 10500:2012** represents the current standard for drinking water quality in India, incorporating both desirable and permissible limits for various parameters. The standard recognizes that achieving desirable limits may not be immediately feasible in all regions, allowing for permissible limits as interim targets.

The Indian standards place particular emphasis on parameters relevant to local conditions, including **fluoride** due to widespread natural occurrence in groundwater, **iron and manganese** from geological sources, and **total dissolved solids (TDS)** reflecting the mineralized nature of many Indian water sources. The desirable limit for TDS is 500 mg/L, with a permissible limit of 2000 mg/L in the absence of alternative sources.

Bacteriological requirements in IS 10500:2012 specify that drinking water should be free from *E. coli* and coliform bacteria throughout the distribution system. The standard also addresses **chlorine residual** requirements for disinfected water supplies, specifying minimum residual chlorine concentrations of 0.2 mg/L at consumer taps to ensure microbiological safety during distribution.

Analytical Techniques: Titration, Colorimetry, Spectrophotometry

Analytical chemistry provides the fundamental tools for water quality assessment, enabling quantitative determination of chemical parameters essential for public health protection and environmental monitoring. The selection of appropriate analytical methods depends on the target analytes, required detection limits, sample matrix effects, and available instrumentation. Classical wet chemistry methods such as titration and colorimetry remain important for routine water analysis, while instrumental techniques including spectrophotometry enable more sensitive and selective determinations.

Titrimetric analysis represents one of the oldest and most reliable approaches for water quality determination, based on the quantitative reaction between the analyte and a standard solution of known concentration (titrant). The **equivalence point**, where stoichiometrically equivalent amounts of analyte and titrant have reacted, is detected through visual or instrumental endpoint detection methods. Titrimetric methods are particularly valuable for water analysis because they provide high precision and accuracy without requiring expensive instrumentation.

Acid-base titrations are extensively used for determining alkalinity, acidity, and pH-related parameters in water samples. **Alkalinity titration** involves the neutralization of bicarbonate, carbonate, and hydroxide ions with standardized hydrochloric acid:

$$HCO_3^- + H^+ \rightarrow H_2CO_3 \rightarrow H_2O + CO_2 \quad (2.49)$$

$$CO_3^{2-} + 2H^+ \rightarrow H_2CO_3 \rightarrow H_2O + CO_2 \quad (2.50)$$

The **methyl orange endpoint** (pH 4.3) corresponds to the complete neutralization of all alkaline species, providing a measure of total alkalinity expressed as mg/L $CaCO_3$. This parameter is crucial for assessing the acid-neutralizing capacity of water and predicting corrosion potential in distribution systems.

Complexometric titrations using ethylenediaminetetraacetic acid (EDTA) provide accurate determinations of water hardness and individual metal concentrations. **Total hardness** is determined through titration with EDTA at pH 10 using Eriochrome Black T indicator:

$$Ca^{2+} + EDTA^{4-} \rightarrow [Ca\text{-}EDTA]^{2-} \quad (2.51)$$

$$Mg^{2+} + EDTA^{4-} \rightarrow [Mg\text{-}EDTA]^{2-} \quad (2.52)$$

The **stability constants** of metal-EDTA complexes are sufficiently large ($\log K > 10$) to ensure quantitative complexation at appropriate pH conditions. **Calcium hardness** can be determined separately using the same titration procedure at pH 12-13, where magnesium is precipitated as $Mg(OH)_2$ and does not interfere with the calcium determination.

Redox titrations are employed for determining oxidizing and reducing species in water samples. **Dissolved oxygen**determination using the **Winkler method** represents a classic redox titration procedure:

$$Mn^{2+} + 2OH^- \rightarrow Mn(OH)_2 \quad (2.53)$$

$$2Mn(OH)_2 + O_2 \rightarrow 2MnO(OH)_2 \quad (2.54)$$

$$MnO(OH)_2 + 2I^- + 4H^+ \rightarrow Mn^{2+} + I_2 + 3H_2O \quad (2.55)$$

$$I_2 + 2S_2O_3^{2-} \rightarrow 2I^- + S_4O_6^{2-} \quad (2.56)$$

This method provides excellent precision (±0.1 mg/L) for dissolved oxygen concentrations and serves as a reference method for calibrating instrumental techniques.

Colorimetric analysis relies on the measurement of light absorption by colored solutions to quantify analyte concentrations. The relationship between absorbance and concentration follows **Beer's Law**:

$$A = \varepsilon bc \quad (2.57)$$

where A is absorbance, ε is the molar absorptivity coefficient, b is the path length, and c is the concentration. This linear relationship enables quantitative analysis through calibration with standard solutions of known concentration.

Spectrophotometric methods for water analysis typically involve the formation of colored complexes between the target analyte and specific reagents. **Orthophosphate determination** using the **molybdenum blue method** exemplifies this approach:

$$PO_4^{3-} + 12MoO_4^{2-} + 24H^+ \rightarrow H_3PO_4 \cdot 12MoO_3 + 12H_2O \qquad (2.58)$$

The phosphomolybdic acid complex is subsequently reduced to form molybdenum blue, which exhibits strong absorption at 880 nm. This method provides detection limits below 0.01 mg/L P, making it suitable for monitoring eutrophication-related phosphorus levels in surface waters.

Nitrate and nitrite analysis employs different colorimetric approaches depending on the required sensitivity and potential interferences. The **cadmium reduction method** for nitrate involves reduction to nitrite followed by diazotization and coupling reactions:

$$NO_3^- + Cd \rightarrow NO_2^- + Cd^{2+} \qquad (2.59)$$

The resulting nitrite reacts with sulfanilamide under acidic conditions to form a diazonium compound, which couples with N-(1-naphthyl)ethylenediamine to produce an azo dye with maximum absorption at 543 nm.

Iron determination in water samples utilizes the **1,10-phenanthroline method**, where ferrous iron forms a stable, intensely colored complex:

$$Fe^{2+} + 3(1,10\text{-phenanthroline}) \rightarrow [Fe(phen)_3]^{2+} \quad (2.60)$$

Total iron analysis requires reduction of ferric iron to ferrous iron using hydroxylamine hydrochloride or ascorbic acid. The orange-red complex exhibits maximum absorption at 510 nm and follows Beer's Law over a wide concentration range.

Chlorine residual analysis is critical for monitoring disinfection effectiveness in treated water systems. The **DPD (N,N-diethyl-p-phenylenediamine) colorimetric method** provides rapid and reliable determination of free chlorine, combined chlorine, and total chlorine:

$$DPD + HOCl \rightarrow DPD^{+\bullet} + Cl^- + OH^- \quad (2.61)$$

The DPD radical cation exhibits a magenta color with maximum absorption at 515 nm. **Free chlorine** reacts instantaneously with DPD, while **combined chlorine** (chloramines) requires the addition of potassium iodide to catalyze the reaction.

Spectrophotometric instrumentation for water analysis ranges from simple filter photometers to sophisticated double-beam UV-visible spectrophotometers. **Single-beam instruments** require careful attention to baseline stability and thermal equilibration, while **double-beam instruments** automatically compensate for source intensity variations and provide superior precision for quantitative analysis.

Quality control in water analysis requires attention to sampling procedures, sample preservation, analytical blanks, and calibration verification. **Analytical blanks** prepared with reagent-grade water help identify contamination from reagents or glassware, while **matrix spikes** assess potential interferences from sample components. **Duplicate analyses** provide measures of analytical precision, while **certified reference materials** enable accuracy verification.

2.3 Water Quality Parameters and Public Health Relevance

Determination of BOD, COD, Nitrate, Phosphate, Hardness

Biochemical Oxygen Demand (BOD) represents one of the most important parameters for assessing organic pollution in aquatic systems and the efficiency of wastewater treatment processes. BOD measures the amount of dissolved oxygen consumed by microorganisms during the aerobic decomposition of organic matter over a specified time period, typically five days (BOD_5). This parameter provides a direct indication of the oxygen-depleting potential of wastewater and its impact on receiving water bodies.

The **BOD test procedure** involves incubating water samples in sealed bottles at 20°C in the dark for five days, measuring the decrease in dissolved oxygen concentration. The samples must be diluted to ensure that dissolved oxygen consumption is neither complete nor negligible, typically targeting 40-70% oxygen depletion. **Seed microorganisms** are added to samples with low microbial populations to ensure adequate biological activity for organic matter decomposition.

The **theoretical basis** of BOD relates to the aerobic respiration of organic matter by microorganisms:

$$\text{Organic matter} + O_2 + \text{microorganisms} \rightarrow CO_2 + H_2O + \text{biomass} \quad (2.62)$$

The oxygen consumption follows first-order kinetics, described by the equation:

$$BOD_t = BOD\_ultimate(1 - e^{(-kt)}) \quad (2.63)$$

where BOD_t is the oxygen demand at time t, BOD_ultimate is the ultimate oxygen demand, and k is the deoxygenation rate constant. For most domestic wastewaters, approximately 68% of the ultimate BOD is exerted in five days when $k = 0.23\ day^{-1}$ at 20°C.

Nitrification interference in BOD measurements occurs when ammonia-oxidizing bacteria consume oxygen through nitrification reactions:

$$NH_4^+ + 1.5O_2 \rightarrow NO_2^- + 2H^+ + H_2O \quad (2.64)$$

$$NO_2^- + 0.5O_2 \rightarrow NO_3^- \quad (2.65)$$

This interference is prevented by adding nitrification inhibitors such as allylthiourea (ATU) or by maintaining samples at low temperature to suppress nitrifying bacteria growth.

Chemical Oxygen Demand (COD) provides a measure of the total organic and inorganic matter that can be oxidized by a strong chemical oxidant, typically potassium dichromate in sulfuric acid solution. COD analysis is completed in approximately 2 hours compared to 5 days for BOD, making it valuable for process control and monitoring applications. The COD test oxidizes both biodegradable and non-biodegradable organic compounds, typically yielding higher values than BOD.

The **COD reaction mechanism** involves the oxidation of organic compounds by dichromate in acidic solution:

$$Cr_2O_7^{2-} + 14H^+ + 6e^- \rightarrow 2Cr^{3+} + 7H_2O \quad (2.66)$$

For a generalized organic compound $C_aH_\beta Oc$, the oxidation reaction is:

$$C_aH_\beta Oc + (2a + b/2 - c)/3\ Cr_2O_7^{2-} + (8a + 2b - 4c)/3\ H^+ \rightarrow$$

$$aCO_2 + (b + 8a + 2b - 4c)/6\ H_2O + (2a + b/2 - c)/3 \times 2Cr^{3+} \quad (2.67)$$

The **COD procedure** involves heating samples with excess dichromate and sulfuric acid at 150°C for 2 hours in sealed vials containing silver catalyst. The residual dichromate is measured spectrophotometrically at 600 nm, or the produced chromic ion is measured at 420 nm. **Chloride interference** is significant in COD analysis, as chloride ions are oxidized by dichromate, leading to elevated COD values in saline waters.

BOD/COD ratios provide insights into the biodegradability of organic matter in water samples. Ratios of 0.4-0.8 indicate readily biodegradable organic matter typical of domestic wastewater, while ratios below 0.3 suggest the presence of refractory organic compounds that resist biological decomposition. Industrial wastewaters often exhibit low BOD/COD ratios due to the presence of synthetic organic compounds.

Nitrate determination is essential for assessing eutrophication potential, drinking water safety, and groundwater contamination from agricultural sources. **Ultraviolet absorption** provides a rapid screening method for nitrate based on absorption at 220 nm:

$$A_{220} = \varepsilon_{220} \times c \times l \quad (2.68)$$

where ε_{220} is the molar absorptivity of nitrate at 220 nm (approximately 7 L mol^{-1} cm^{-1}). **Organic matter interference** is corrected by measuring absorption at 275 nm and applying a correction factor, as organic compounds typically exhibit absorption at both wavelengths while nitrate shows minimal absorption at 275 nm.

Ion chromatography provides the most reliable method for nitrate analysis in complex matrices, utilizing ion-exchange separation followed by conductivity detection. **Anion-exchange columns** separate nitrate from other anions based on their affinity for the stationary phase, with typical elution using carbonate/bicarbonate eluents. Detection limits of 0.01 mg/L N are routinely achieved, making this method suitable for monitoring low-level contamination.

Phosphate analysis is critical for eutrophication assessment, as phosphorus is typically the limiting nutrient in freshwater systems. **Orthophosphate** represents the most readily available form for algal uptake, while **total phosphorus** includes organic phosphorus compounds and polyphosphates that can be hydrolyzed to orthophosphate.

The **molybdenum blue method** for orthophosphate involves the formation of a phosphomolybdic acid complex that is subsequently reduced to form intensely colored molybdenum blue:

$$PO_4^{3-} + 12MoO_4^{2-} + 24H^+ \rightarrow H_3[P(Mo_3O_{10})_4] + 12H_2O \quad (2.69)$$

Ascorbic acid reduction of the phosphomolybdic acid complex produces molybdenum blue with maximum absorption at 880 nm. This method provides detection limits of approximately 0.01 mg/L P and exhibits minimal interference from other common water constituents.

Total phosphorus analysis requires **acid persulfate digestion** to convert organic phosphorus compounds and polyphosphates to orthophosphate:

$$\text{Organic P compounds} + S_2O_8^{2-} + H^+ \rightarrow PO_4^{3-} + SO_4^{2-} + \ldots \quad (2.70)$$

The digestion is typically performed in an autoclave at 121°C for 30 minutes or by heating at 150°C for 30 minutes, followed by orthophosphate analysis using the molybdenum blue method.

Water hardness results primarily from dissolved calcium and magnesium ions acquired through contact with limestone, dolomite, and other geological formations. Hardness affects soap consumption, scale formation in pipes and appliances, and aesthetic qualities of water. The **classification of water hardness** ranges from soft (< 75 mg/L as $CaCO_3$) to very hard (> 300 mg/L as $CaCO_3$).

EDTA titration remains the standard method for hardness determination, based on the formation of stable 1:1 complexes between EDTA and divalent cations:

$$Ca^{2+} + H_2Y^{2-} \rightarrow CaY^{2-} + 2H^+ \tag{2.71}$$

$$Mg^{2+} + H_2Y^{2-} \rightarrow MgY^{2-} + 2H^+ \tag{2.72}$$

where H_2Y^{2-} represents the predominant EDTA species at pH 10. **Eriochrome Black T** serves as the endpoint indicator, forming wine-red complexes with calcium and magnesium that are displaced by EDTA to produce a blue endpoint.

Calcium hardness is determined separately using the same EDTA titration at pH 12-13, where magnesium precipitates as $Mg(OH)_2$ and does not interfere. **Magnesium hardness** is calculated as the difference between total hardness and calcium hardness.

Methods: Winkler's Method, UV-vis Absorption, Ion Chromatography

Winkler's method for dissolved oxygen determination represents one of the most precise and accurate analytical procedures in environmental chemistry, serving as the reference method for calibrating instrumental oxygen analyzers. This iodometric titration procedure, developed by Lajos Winkler in 1888, provides precision better than ±0.1 mg/L when performed carefully, making it suitable for critical applications such as BOD analysis and water quality monitoring.

The **chemical principles** underlying Winkler's method involve the quantitative oxidation of manganous hydroxide by dissolved oxygen to form higher oxidation state manganese compounds:

$$Mn^{2+} + 2OH^- \rightarrow Mn(OH)_2 \tag{2.73}$$

$$2Mn(OH)_2 + O_2 \rightarrow 2MnO(OH)_2 \quad (2.74)$$

Under acidic conditions in the presence of iodide, the manganic compounds oxidize iodide to iodine stoichiometrically:

$$MnO(OH)_2 + 2I^- + 4H^+ \rightarrow Mn^{2+} + I_2 + 3H_2O \quad (2.75)$$

The liberated iodine is then titrated with standardized sodium thiosulfate:

$$I_2 + 2S_2O_3^{2-} \rightarrow 2I^- + S_4O_6^{2-} \quad (2.76)$$

Procedural considerations are critical for accurate Winkler determinations. **Sample collection** must avoid air entrainment, requiring careful filling of BOD bottles with overflow to eliminate bubbles. **Reagent addition** follows a specific sequence: manganous sulfate solution is added first, followed immediately by alkaline potassium iodide solution. The sample must be mixed thoroughly after each reagent addition to ensure complete precipitation of manganous hydroxide.

Interference corrections address the presence of oxidizing and reducing substances that affect the iodine-thiosulfate stoichiometry. **Nitrite interference** is common in polluted waters and wastewater:

$$NO_2^- + I^- + 2H^+ \rightarrow NO + I_2 + H_2O \quad (2.77)$$

This interference is corrected by adding sodium azide, which destroys nitrite:

$$N_3^- + NO_2^- + 2H^+ \rightarrow N_2O + N_2 + H_2O \quad (2.78)$$

Ferrous iron interference occurs in anoxic waters containing dissolved iron:

$$2Fe^{3+} + 2I^- \rightarrow 2Fe^{2+} + I_2 \quad (2.79)$$

This positive interference (leading to erroneously high dissolved oxygen values) is addressed by adding potassium fluoride to complex ferric iron and prevent iodine liberation.

Temperature and salinity corrections are essential for accurate dissolved oxygen determination, as oxygen solubility varies significantly with these parameters. The **Henry's Law constant** for oxygen dissolution decreases with increasing temperature, reducing equilibrium oxygen concentrations:

$$C = K\_H \times P\_O_2 \quad (2.80)$$

where C is the dissolved oxygen concentration, K_H is Henry's constant, and P_O_2 is the partial pressure of oxygen. **Salinity effects** result from the "salting out" of dissolved gases by ionic species, with approximately 20% reduction in oxygen solubility in seawater compared to freshwater at the same temperature.

UV-visible absorption spectrophotometry provides versatile and sensitive methods for numerous water quality parameters. The **theoretical foundation** rests on Beer's Law, which relates light absorption to the concentration of absorbing species:

$$A = \log(I_0/I) = \varepsilon bc \quad (2.81)$$

where A is absorbance, I_0 and I are incident and transmitted light intensities, ε is the molar absorptivity, b is the path length, and c is the concentration.

Instrumentation considerations significantly affect analytical performance. **Double-beam spectrophotometers**automatically compensate for source intensity fluctuations and cell differences by measuring sample and reference beams simultaneously. **Wavelength accuracy** and **spectral bandwidth** influence method sensitivity and selectivity, with narrower bandwidths generally providing better resolution but reduced light throughput.

Matrix effects in UV-visible analysis can arise from light scattering by turbidity, absorption by interfering species, and changes in solution chemistry that affect chromophore formation. **Turbidity compensation** often requires filtration or turbidity-corrected measurements at non-absorbing wavelengths. **Background correction** procedures subtract absorption by sample matrices from total absorption to improve accuracy.

Method validation for UV-visible procedures requires assessment of linearity, precision, accuracy, detection limits, and interference effects. **Calibration curves** should encompass the expected concentration range and demonstrate linearity with correlation coefficients exceeding 0.995. **Detection limits** are typically calculated as three times the standard deviation of blank measurements divided by the slope of the calibration curve.

Ion chromatography (IC) has revolutionized the analysis of ionic species in water samples, providing simultaneous determination of multiple anions or cations with excellent sensitivity and selectivity. The **separation mechanism** relies on differential affinity of ions for ion-exchange sites on the stationary phase, typically cross-linked polystyrene resins functionalized with quaternary ammonium groups for anion analysis or sulfonic acid groups for cation analysis.

Anion chromatography commonly employs carbonate/bicarbonate eluents for the separation of common anions including fluoride, chloride, nitrite, bromide, nitrate, phosphate, and sulfate. The **retention mechanism** involves competition between analyte anions and eluent anions for positively charged sites on the stationary phase:

$$R\text{-}N^+(CO_3^{2-}) + A^- \leftrightarrow R\text{-}N^+(A^-) + CO_3^{2-} \qquad (2.82)$$

Gradient elution programs optimize separation by increasing eluent strength during the chromatographic run, allowing rapid elution of strongly retained anions while maintaining resolution for weakly retained species.

Suppressed conductivity detection represents the most common detection method for IC, utilizing ion-exchange membranes to convert eluent ions to weakly conductive species while converting analyte ions to highly conductive acids or bases. For anion chromatography with carbonate eluent, the suppressor converts carbonate to carbonic acid:

$$CO_3^{2-} + 2H^+ \rightarrow H_2CO_3 \qquad (2.83)$$

Analyte anions are converted to their corresponding acids:

$$A^- + H^+ \rightarrow HA \qquad (2.84)$$

Quality assurance in IC analysis requires attention to column conditioning, eluent preparation, and system contamination. **Eluent degassing** prevents bubble formation that can affect baseline stability and peak resolution. **System contamination** by trace ionic species can elevate detection limits and create ghost peaks, necessitating thorough cleaning procedures and high-purity reagents.

Interpretation of Water Analysis Reports

Water analysis reports provide comprehensive chemical characterization of water samples for various purposes including drinking water compliance, environmental monitoring, and process control. The **effective interpretation** of these reports requires understanding of analytical methods, regulatory standards, sampling procedures, and the relationships between different parameters. Professional interpretation considers not only individual parameter values but also chemical consistency, potential error sources, and environmental or health significance.

Quality control indicators provide the first assessment of data reliability. **Analytical blanks** should show values below detection limits for most parameters, with elevated blank values indicating potential contamination during sampling, preservation, or analysis. **Duplicate analyses** assess analytical precision, with relative percent differences (RPD) typically required to be less than 20% for most parameters:

$$RPD = |C_1 - C_2|/((C_1 + C_2)/2) \times 100 \qquad (2.85)$$

where C_1 and C_2 are duplicate concentrations.

Ion balance calculations provide fundamental checks of analytical quality for complete water analyses. The **electroneutrality principle** requires that the sum of positive charges (cations) equals the sum of negative charges (anions) in any aqueous solution:

$$\Sigma(\text{cations}) = \Sigma(\text{anions}) \qquad (2.86)$$

Charge balance errors exceeding ±5% suggest analytical problems, incomplete analyses, or the presence of unmeasured ionic species. Common causes include failure to analyze for significant ions, analytical errors, or sample contamination.

Total dissolved solids (TDS) verification compares measured TDS with calculated TDS from individual ion concentrations. The **calculated**

TDS includes all major ions weighted by their relative atomic or molecular masses:

$$TDS(calc) = Ca^{2+} + Mg^{2+} + Na^{+} + K^{+} + HCO_3^{-} + SO_4^{2-} + Cl^{-} + NO_3^{-} \quad (2.87)$$

Significant discrepancies between measured and calculated TDS may indicate the presence of unmeasured dissolved species or analytical errors.

Hardness relationships provide additional data quality checks. **Total hardness** should approximately equal the sum of calcium and magnesium hardness:

$$\text{Total hardness} \approx \text{Ca hardness} + \text{Mg hardness} \quad (2.88)$$

Calculated hardness from individual calcium and magnesium concentrations should agree with titrated hardness within analytical uncertainty.

Regulatory compliance assessment requires comparison of measured concentrations with applicable standards or guidelines. **Drinking water standards** are typically expressed as maximum contaminant levels (MCLs) that cannot be exceeded, while **secondary standards** address aesthetic qualities such as taste, odor, and color. **Trend analysis** of multiple sampling events provides insights into temporal variations and potential contamination sources.

Health risk evaluation considers both acute and chronic exposure scenarios. **Acute risks** relate to immediate health effects from short-term exposure to high contaminant concentrations, while **chronic risks** involve long-term exposure to lower concentrations. **Vulnerable populations** including infants, pregnant women, and immunocompromised individuals may face elevated risks from contaminants that pose minimal risks to healthy adults.

Environmental significance of water quality data depends on the receiving water body characteristics and intended uses. **Surface water quality** affects aquatic ecosystem health, with parameters such as dissolved oxygen, pH, and nutrient concentrations directly influencing biological communities. **Groundwater quality** considerations include natural background concentrations, contamination source identification, and potential impacts on wells and springs.

Seasonal variations in water quality parameters reflect natural processes and anthropogenic activities. **Spring snowmelt**often increases turbidity and decreases concentrations of dissolved constituents through dilution effects. **Summer conditions** may promote algal growth in surface waters, affecting dissolved oxygen, pH, and nutrient concentrations. **Agricultural seasons** influence nitrate and pesticide concentrations in both surface and groundwater.

Geochemical indicators provide insights into water-rock interactions and residence times. **Sodium-chloride ratios** help distinguish between freshwater and saltwater influences, with marine ratios around 0.86 and freshwater ratios typically much higher. **Calcium-magnesium ratios** reflect the relative dissolution of limestone (high Ca/Mg) versus dolomite (lower Ca/Mg).

Contamination source identification utilizes chemical fingerprints to distinguish between different pollution sources. **Nitrate-to-chloride ratios** help differentiate between fertilizer contamination (high NO_3/Cl) and sewage contamination (lower NO_3/Cl ratios). **Stable isotope ratios** of nitrogen and oxygen in nitrate provide additional source identification capabilities.

Corrosion indices assess the aggressiveness of water toward metal pipes and plumbing fixtures. The **Langelier Saturation Index (LSI)** indicates the tendency for calcium carbonate precipitation or dissolution:

$$LSI = pH - pH\_s \quad (2.89)$$

where pH_s is the pH at calcium carbonate saturation. Positive LSI values indicate scale-forming tendency, while negative values suggest corrosive conditions.

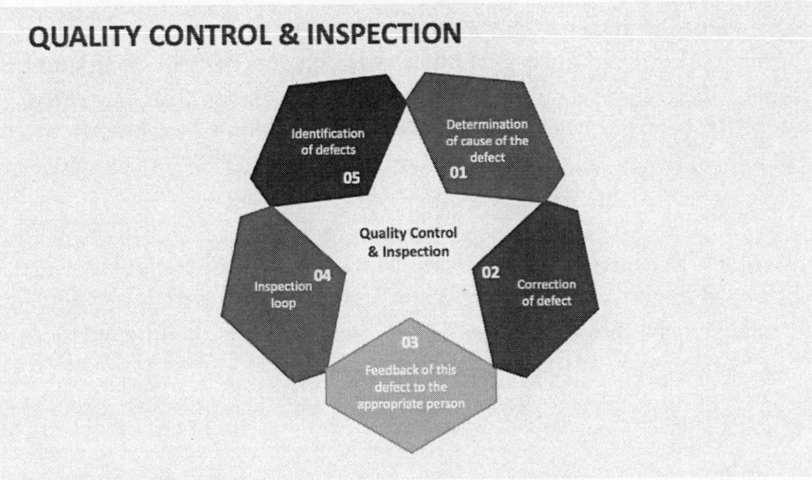

Fig. 9: Flowchart for systematic interpretation of water analysis reports showing quality control checks, regulatory comparisons, and environmental significance assessment

2.4 Effluent and Wastewater Treatment Techniques

Primary, Secondary, and Tertiary Treatment Processes

Wastewater treatment represents a critical application of chemical and biological principles to protect public health and environmental quality. The **hierarchical treatment approach** involves progressively more sophisticated processes to remove different categories of contaminants. **Primary treatment** focuses on physical separation of settleable and floatable materials, **secondary treatment** employs biological processes to remove dissolved and colloidal organic matter, and **tertiary treatment** addresses remaining contaminants including nutrients, pathogens, and trace pollutants.

Primary treatment processes utilize **physical separation mechanisms** to remove approximately 25-50% of biochemical oxygen demand (BOD) and 50-70% of total suspended solids (TSS) from wastewater. **Screening** represents the initial treatment step, employing bar screens with 6-25 mm openings to remove large debris, rags, and other materials that could damage downstream equipment or interfere with treatment processes.

Grit removal systems eliminate sand, gravel, and other dense inorganic particles that can cause excessive wear on mechanical equipment and reduce the effective volume of biological treatment units. **Aerated grit chambers** introduce air to maintain organic matter in suspension while allowing denser grit particles to settle, with typical detention times of 2-5 minutes and horizontal velocities of 0.25-0.40 m/s.

Primary sedimentation occurs in circular or rectangular clarifiers designed to provide quiescent conditions for the gravitational settling of suspended particles. **Stokes' Law** governs the settling velocity of discrete particles in viscous flow:

$$v_s = (\rho_p - \rho_w)gd^2/18\mu \quad (2.90)$$

where v_s is settling velocity, ρ_p and ρ_w are particle and water densities, g is gravitational acceleration, d is particle diameter, and μ is dynamic viscosity.

Overflow rates (surface loading rates) typically range from 30-50 m^3/m^2/day for primary clarifiers, with hydraulic retention times of 1.5-2.5 hours. **Zone settling** occurs for flocculent particles that aggregate during settling, creating distinct settling zones with different velocities. **Primary sludge production** typically ranges from 0.8-1.2 kg TSS per kg BOD removed, with moisture content around 95-97%.

Chemical precipitation can enhance primary treatment through the addition of coagulants such as ferric chloride, aluminumsulfate, or lime. **Coagulation mechanisms** involve charge neutralization and sweep flocculation:

$$Al_2(SO_4)_3 + 6H_2O \rightarrow 2Al(OH)_3 + 3H_2SO_4 \quad (2.91)$$

The aluminum hydroxide precipitate enmeshes suspended particles and provides sites for adsorption of dissolved phosphorus compounds.

Secondary treatment systems employ **biological processes** to remove dissolved and colloidal organic matter through microbial metabolism. These systems rely on mixed microbial communities that oxidize organic substrates using molecular oxygen (aerobic processes) or alternative electron acceptors (anoxic/anaerobic processes). Secondary treatment typically achieves 85-95% BOD removal and 85-92% TSS removal.

Activated sludge processes represent the most widely used aerobic biological treatment technology, utilizing suspended microbial biomass to metabolize organic pollutants. The **fundamental biochemical reactions** involve the oxidation of organic matter and the synthesis of new biomass:

$$\text{Organic matter} + O_2 + \text{nutrients} \rightarrow \text{new cells} + CO_2 + H_2O + \text{energy} \quad (2.92)$$

Microbial kinetics in activated sludge systems follow Monod relationships for substrate utilization:

$$\mu = \mu\_max \times S/(K\_s + S) \quad (2.93)$$

where μ is specific growth rate, μ_max is maximum specific growth rate, S is substrate concentration, and K_s is the half-saturation constant.

Mixed liquor suspended solids (MLSS) concentrations typically range from 2,000-4,000 mg/L in conventional activated sludge systems, with **mixed liquor volatile suspended solids (MLVSS)** representing approximately 70-80% of MLSS. The **food-to-microorganism ratio (F/M)** provides operational control:

$$F/M = BOD_5 \text{ loading}/(MLVSS \times \text{aeration volume}) \quad (2.94)$$

Typical F/M ratios range from 0.2-0.5 kg BOD_5/kg MLVSS/day for conventional systems.

Oxygen requirements for activated sludge processes include both carbonaceous oxygen demand (BOD removal) and nitrogenous oxygen demand (ammonia oxidation). **Theoretical oxygen requirements** can be calculated from stoichiometric relationships:

$$C_6H_{12}O_6 + 6O_2 \rightarrow 6CO_2 + 6H_2O \quad (2.95)$$

$$\text{For nitrification: } NH_4^+ + 2O_2 \rightarrow NO_3^- + 2H^+ + H_2O \quad (2.96)$$

Aeration systems maintain dissolved oxygen concentrations of 1-3 mg/L in activated sludge reactors through mechanical surface aerators or diffused air systems. **Oxygen transfer efficiency** depends on factors including bubble size, contact time, temperature, and wastewater characteristics.

Secondary clarification separates treated effluent from biological sludge, with design considerations including **state point analysis** to prevent sludge blanket rise and effluent quality deterioration. **Solids loading rates** typically range from 100-150 kg/m²/day, with hydraulic overflow rates of 16-24 m³/m²/day.

Return activated sludge (RAS) maintains appropriate biomass concentrations in the aeration basin, with return ratios typically 25-100% of influent flow. **Waste activated sludge (WAS)** controls the **solids retention time (SRT)**:

$$SRT = MLSS \times \text{aeration volume/WAS flow rate} \tag{2.97}$$

Typical SRT values range from 5-15 days for conventional activated sludge systems.

Tertiary treatment processes address contaminants not adequately removed by conventional primary and secondary treatment, including nutrients (nitrogen and phosphorus), pathogens, trace organics, and residual suspended solids. **Advanced treatment requirements** are driven by stringent discharge standards, water reuse applications, and protection of sensitive receiving waters.

Biological nutrient removal (BNR) systems modify conventional activated sludge processes to achieve nitrogen and phosphorus removal through controlled redox conditions. **Nitrification** converts ammonia to nitrate under aerobic conditions with specialized autotrophic bacteria:

$$NH_4^+ + 1.5O_2 \rightarrow NO_2^- + 2H^+ + H_2O \tag{2.98}$$

$$NO_2^- + 0.5O_2 \rightarrow NO_3^- \tag{2.99}$$

Denitrification reduces nitrate to nitrogen gas under anoxic conditions with heterotrophic bacteria using organic carbon as electron donor:

$$5C_6H_{12}O_6 + 24NO_3^- \rightarrow 30CO_2 + 18H_2O + 12N_2 + 24OH^- \tag{2.100}$$

Enhanced biological phosphorus removal (EBPR) utilizes polyphosphate-accumulating organisms (PAOs) that store phosphorus as polyphosphate under aerobic conditions and release it under anaerobic conditions. The process requires alternating anaerobic and aerobic zones to select for PAOs.

Chemical phosphorus removal employs metal salts to precipitate phosphorus as insoluble compounds:

$$Al^{3+} + PO_4^{3-} \rightarrow AlPO_4 \qquad (2.101)$$

$$Fe^{3+} + PO_4^{3-} \rightarrow FePO_4 \qquad (2.102)$$

Filtration systems remove residual suspended solids and associated contaminants through various mechanisms including straining, sedimentation, interception, and adsorption. **Granular media filters** using sand, anthracite, or multimedia typically achieve effluent TSS concentrations below 10 mg/L.

Membrane bioreactors (MBRs) combine biological treatment with membrane filtration, achieving high-quality effluent suitable for water reuse applications. **Microfiltration and ultrafiltration membranes** with pore sizes of 0.01-0.1 μm effectively remove bacteria, viruses, and suspended solids.

Activated Sludge Process, Trickling Filters

The activated sludge process represents the most widely implemented biological wastewater treatment technology globally, treating millions of cubic meters of municipal and industrial wastewater daily. This **suspended growth system**maintains high concentrations of microorganisms in an aerated reactor where organic pollutants are metabolized and converted to carbon dioxide, water, and new biomass. The process effectiveness relies on maintaining optimal environmental conditions for diverse microbial communities capable of degrading various organic compounds.

Process configuration variations have evolved to address specific treatment objectives and operational constraints. **Conventional activated sludge** employs plug-flow hydraulics with uniform air distribution, providing hydraulic retention times of 4-8 hours and SRT values of 5-15 days. **Complete-mix activated sludge** provides uniform conditions throughout the reactor, offering advantages for variable loading conditions and inhibitory wastes.

Extended aeration systems operate at low organic loading rates (0.05-0.15 kg BOD_5/kg MLVSS/day) with SRT values of 20-30 days, promoting complete stabilization of organic matter and reducing excess sludge production. These systems are particularly suitable for small communities due to their operational simplicity and process stability.

Step-feed activated sludge introduces wastewater at multiple points along the reactor length, distributing organic loading and improving oxygen utilization efficiency. This configuration helps prevent oxygen depletion and provides more uniform loading on secondary clarifiers.

Microbial ecology in activated sludge systems involves complex communities including bacteria, protozoa, metazoa, and fungi. **Heterotrophic bacteria** constitute the dominant group responsible for organic matter removal, with genera such as *Pseudomonas*, *Bacillus*, *Acinetobacter*, and *Alcaligenes* commonly identified. **Filamentous bacteria** can cause sludge bulking problems when their growth is favored by low dissolved oxygen, high F/M ratios, or nutrient deficiencies.

Protozoa populations serve as indicators of process performance and effluent quality. **Free-swimming ciliates** typically dominate in young, high-loaded systems, while **stalked ciliates** and **crawling ciliates** are associated with well-nitrified, stable systems producing high-quality effluent. **Amoebae** often indicate poor settling sludge and marginal effluent quality.

Process control parameters require continuous monitoring to maintain optimal performance. **Dissolved oxygen concentrations** should be maintained at 1-3 mg/L to support aerobic metabolism while avoiding excessive energy consumption. **pH control** typically maintains values between 6.5-8.5, with biological buffering capacity usually adequate for municipal wastewaters.

Sludge volume index (SVI) provides a measure of sludge settleability:

$$SVI = (\text{settled sludge volume}/MLSS) \times 1000 \qquad (2.103)$$

Good settling sludge typically exhibits SVI values of 80-120 mL/g, while values above 150 mL/g indicate poor settleability and potential clarifier problems.

Nutrient requirements for activated sludge processes follow approximate ratios of BOD_5:N:P = 100:5:1 for municipal wastewaters. **Nitrogen deficiency** can limit biological activity and promote filamentous growth, while **phosphorus deficiency** affects energy metabolism and cell synthesis.

Process variations address specific treatment objectives through modified operational strategies. **Oxidation ditches**employ extended aeration in ring-shaped reactors with mechanical surface aerators or brush rotors, providing simple operation and good treatment performance for small to medium-sized communities.

Sequencing batch reactors (SBRs) conduct all treatment steps in a single reactor through timed cycles of fill, react, settle, and decant phases. This configuration offers operational flexibility and can achieve biological nutrient removal through controlled aerobic and anoxic conditions.

Trickling filter systems represent **attached growth biological treatment** processes where microorganisms grow as biofilms on inert support media. Wastewater is distributed over the media surface and trickles downward while air flows upward through void spaces, providing oxygen for aerobic metabolism. These systems offer advantages including low energy requirements, simple operation, and resistance to shock loads and toxic materials.

Media characteristics significantly affect trickling filter performance. **Natural rock media** with sizes of 25-100 mm provides high void ratios (45-55%) and good structural stability but exhibits low specific surface area (40-60 m^2/m^3). **Synthetic plastic media** offers higher specific surface areas (80-200 m^2/m^3) and lighter weight, enabling deeper filter construction and higher hydraulic loading rates.

Biofilm development occurs through the attachment and growth of microorganisms on media surfaces. **Primary colonizers** including bacteria and fungi establish initial biofilm layers, followed by higher organisms such as protozoa and metazoa. **Biofilm thickness** typically ranges from 0.1-2.0 mm, with older biofilm layers becoming anaerobic due to oxygen diffusion limitations.

Mass transfer processes in trickling filters involve the transport of substrates and oxygen from the bulk liquid to the biofilm, and the removal of metabolic products. **Substrate removal follows zero-order kinetics** in thick biofilms where substrate concentrations at the biofilm surface are high:

$dS/dt = -k_0$ (2.104)

where k_0 is the zero-order rate constant.

Hydraulic loading rates typically range from 10-40 $m^3/m^2/day$ for standard-rate trickling filters and 14-185 $m^3/m^2/day$ for high-rate systems. **Organic loading rates** range from 0.08-0.32 kg $BOD_5/m^3/day$ for standard-rate filters and 0.32-1.12 kg $BOD_5/m^3/day$ for high-rate systems.

Recirculation is commonly employed in trickling filter systems to dilute influent strength, maintain adequate hydraulic loading during low-flow periods, and improve treatment efficiency. **Recirculation ratios** typically range from 0.5-3.0 times the average daily flow, calculated as:

R = recirculation flow/influent flow (2.105)

Temperature effects significantly influence trickling filter performance, with activity decreasing substantially below 10°C and increasing with temperature up to approximately 35°C. **Winter operation** in cold climates may require insulation or housing to maintain treatment effectiveness.

High-rate trickling filters operate at higher hydraulic and organic loading rates, typically requiring secondary clarification and often employing intermediate clarification with sludge return. These systems can achieve BOD_5 removal efficiencies of 65-85% compared to 80-95% for standard-rate systems.

Rotating biological contactors (RBCs) represent a variation of attached growth treatment utilizing plastic discs that rotate through wastewater and air, providing alternating conditions for biofilm development. **Disc rotation speeds** of 1-2 rpm maintain adequate mixing and biofilm shearing while minimizing energy requirements.

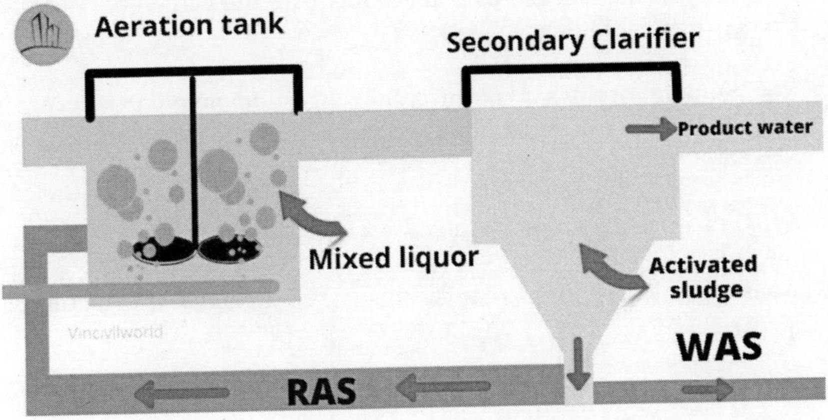

Fig. 10: Cross-sectional diagram of activated sludge process showing aeration basin, secondary clarifier, and sludge return/waste systems with typical operating parameters

Sludge Disposal, Oxidation Ponds

Sludge management represents one of the most challenging and expensive aspects of wastewater treatment, typically accounting for 25-50% of total treatment plant operating costs. **Sewage sludge characteristics** vary significantly depending on the source wastewater composition and treatment processes employed, with typical solids concentrations ranging from 0.25-12% depending on the degree of thickening and dewatering applied.

Primary sludge from municipal wastewater treatment exhibits high organic content (60-80% volatile solids), putrescible nature, and high pathogen concentrations. **Waste activated sludge** contains higher concentrations of nitrogen and phosphorus due to microbial cell composition but generally has better dewatering characteristics than primary sludge. **Mixed sludges** combining primary and secondary sources require careful management to optimize treatment processes.

Sludge stabilization processes reduce vector attraction, pathogen concentrations, and odor generation while improving handling characteristics. **Anaerobic digestion** represents the most widely used stabilization method for municipal sludges, utilizing methanogenic bacteria to convert organic matter to methane and carbon dioxide under oxygen-free conditions.

Biochemical pathways in anaerobic digestion involve three distinct phases. **Hydrolysis** breaks down complex organic polymers to simpler compounds:

$$(C_6H_{10}O_5)_n + nH_2O \rightarrow nC_6H_{12}O_6 \quad (2.106)$$

Acidogenesis converts simple organic compounds to volatile fatty acids:

$$C_6H_{12}O_6 \rightarrow 3CH_3COOH \quad (2.107)$$

Methanogenesis produces methane from acetate and hydrogen/carbon dioxide:

$$CH_3COOH \rightarrow CH_4 + CO_2 \quad (2.108)$$

$$4H_2 + CO_2 \rightarrow CH_4 + 2H_2O \quad (2.109)$$

Process design parameters for anaerobic digestion include hydraulic retention times of 15-20 days at mesophilic temperatures (35°C) or 12-14 days at thermophilic temperatures (55°C). **Volatile acids to alkalinity ratios** should be maintained below 0.1-0.3 to ensure process stability, with pH values between 6.8-7.2.

Gas production rates typically range from 0.5-1.0 m^3 biogas per kg volatile solids destroyed, with methane content of 60-70%. **Energy recovery** from biogas combustion can offset 50-90% of plant energy requirements, making anaerobic digestion economically attractive for larger treatment facilities.

Aerobic digestion provides an alternative stabilization method particularly suitable for smaller treatment plants. **Open aerobic digestion** maintains dissolved oxygen concentrations above 1 mg/L at temperatures of 15-35°C, with detention times of 40-60 days in cold climates. **Autothermal thermophilic aerobic digestion (ATAD)** operates at temperatures of 40-80°C achieved through exothermic biological reactions.

Lime stabilization raises sludge pH above 12 for at least 2 hours, achieving pathogen reduction through chemical disinfection. **Quicklime (CaO)** or **hydrated lime ($Ca(OH)_2$)** dosages typically range from 50-300 mg/g dry solids depending on sludge buffering capacity and moisture content.

Sludge thickening reduces sludge volume and improves subsequent processing efficiency. **Gravity thickening** achieves solids concentrations of 4-8% for primary sludge and 2-4% for waste activated sludge through quiescent settling in circular tanks with slow-rotating rake mechanisms.

Dissolved air flotation (DAF) thickening is particularly effective for waste activated sludge, utilizing pressurized air dissolved in recycled effluent to generate microscopic bubbles that attach to sludge particles and promote flotation. **Air-to-solids ratios** typically range from 0.005-0.06 kg air/kg solids.

Centrifugal thickening employs high gravitational forces to separate solids from water, achieving solids concentrations of 4-6% for waste activated sludge. **Polymer conditioning** using cationic polyelectrolytes improves thickening efficiency and reduces polymer consumption in subsequent dewatering processes.

Sludge dewatering further reduces moisture content to facilitate handling and disposal. **Belt filter presses** achieve 15-25% solids concentrations through mechanical compression between porous belts, with polymer conditioning essential for floc formation and water release.

Centrifugal dewatering utilizes high-speed centrifuges to separate water from solids, typically achieving 18-25% solids concentrations for municipal sludges. **Scroll centrifuges** provide continuous operation with variable pool depth and differential speed control to optimize separation efficiency.

Filter press dewatering achieves the highest solids concentrations (25-45%) through hydraulic pressure applied to sludge contained between filter cloths. **Chamber filter presses** provide batch operation with cycle times of 2-4 hours including filling, pressing, and cake discharge.

Ultimate disposal options for treated sludge must address regulatory requirements, public acceptance, and long-term sustainability. **Land application** utilizes the beneficial properties of biosolids as soil amendments and fertilizers, providing nitrogen, phosphorus, and organic matter to improve soil structure and fertility.

Class B biosolids meet pathogen reduction requirements through processes including anaerobic digestion, lime stabilization, or composting, and can be applied to agricultural land with site restrictions and crop harvesting limitations. **Class A biosolids** achieve further pathogen reduction through processes including thermophilic digestion, composting, or heat drying, allowing unrestricted land application.

Heavy metal concentrations in biosolids must comply with regulatory limits to prevent soil and groundwater contamination. **Cumulative pollutant loading rates** restrict the total amount of metals that can be applied to land over time, based on soil cation exchange capacity and pH buffering.

Incineration provides volume reduction and pathogen destruction but requires significant energy input and air pollution control equipment. **Multiple hearth furnaces** and **fluidized bed incinerators**operate at temperatures of 760-980°C, with energy recovery potential through steam generation or electricity production.

Oxidation ponds (stabilization ponds) represent one of the most cost-effective wastewater treatment technologies, particularly suitable for small communities in warm climates with available land. These **natural treatment systems**utilize solar energy, atmospheric reaeration, and biological processes to achieve organic matter removal, pathogen reduction, and nutrient transformation.

Facultative ponds represent the most common oxidation pond configuration, with depths of 1.0-2.5 m that create aerobic surface layers and anaerobic bottom layers. **Algal photosynthesis** provides oxygen for aerobic bacteria in surface waters:

$$6CO_2 + 6H_2O + \text{light energy} \rightarrow C_6H_{12}O_6 + 6O_2 \qquad (2.110)$$

Anaerobic bacterial decomposition in bottom sediments converts organic matter to methane, carbon dioxide, and organic acids:

$$C_6H_{12}O_6 \rightarrow 3CO_2 + 3CH_4 \quad (2.111)$$

Design loading rates for facultative ponds typically range from 100-400 kg BOD_5/ha/day in temperate climates and up to 600 kg BOD_5/ha/day in tropical regions. **Detention times** range from 5-30 days depending on climate, loading rates, and effluent quality requirements.

Temperature effects significantly influence oxidation pond performance, with biological activity decreasing dramatically below 10°C and optimal performance occurring at temperatures above 20°C. **Seasonal performance variations** require consideration of winter conditions in temperate climates.

Pond systems configurations often employ multiple ponds in series to optimize treatment performance. **Primary facultative ponds** receive raw wastewater and provide initial treatment, while **secondary ponds** polish effluent quality. **Maturation ponds** with depths of 1.0-1.5 m and detention times of 3-10 days provide final polishing and pathogen removal.

Anaerobic ponds serve as pretreatment units for high-strength wastewaters, operating at depths of 2.5-5.0 m with detention times of 1-7 days. **Organic loading rates** range from 1000-4000 kg BOD_5/ha/day, with methane production providing potential energy recovery opportunities.

Aerated lagoons combine mechanical aeration with extended detention times, providing process control advantages over natural oxidation ponds. **Partial-mix aerated lagoons** maintain suspended solids in suspension through mechanical aeration, while **complete-mix systems** provide uniform conditions throughout the pond volume.

Hydrodynamic considerations affect oxidation pond performance through mixing patterns, dead zones, and short-circuiting. **Baffling systems** improve hydraulic efficiency by reducing short-circuiting and promoting plug-flow conditions. **Wind-induced mixing** provides natural circulation in surface waters, with pond orientation affecting mixing effectiveness.

Detergents and Eutrophication

Eutrophication represents one of the most widespread water quality problems globally, characterized by excessive nutrient enrichment that promotes algal growth and subsequently degrades aquatic ecosystem health. **Cultural eutrophication** accelerated by human activities has transformed many freshwater and marine systems, creating dead zones, harmful algal blooms, and loss of biodiversity. **Detergent contributions** to eutrophication historically involved phosphate builders but continue through various chemical constituents and indirect effects.

Phosphorus biochemistry in aquatic systems determines the eutrophication potential of different chemical forms. **Orthophosphate (PO_4^{3-})** represents the most bioavailable form, directly assimilated by algae and aquatic plants through membrane transport systems. **Polyphosphates** from detergent formulations undergo hydrolysis to orthophosphate through both chemical and enzymatic pathways:

$$P_3O_{10}^{5-} + H_2O \rightarrow PO_4^{3-} + P_2O_7^{4-} \qquad (2.112)$$

$$P_2O_7^{4-} + H_2O \rightarrow 2PO_4^{3-} \qquad (2.113)$$

Organic phosphorus compounds from detergent additives and degradation products undergo mineralization through microbial decomposition, releasing bioavailable phosphorus with time delays that can sustain eutrophication long after external loading reductions.

Historical detergent composition evolved significantly in response to environmental concerns. **Sodium tripolyphosphate (STPP)** served as the primary builder in laundry detergents from the 1940s through 1970s, typically comprising 30-50% of product weight. **Builder functions** include water softening through complexation of hardness ions, suspension of soil particles, and pH buffering:

$$Ca^{2+} + P_3O_{10}^{5-} \rightarrow [Ca\text{-}P_3O_{10}]^{3-} \qquad (2.114)$$

Phosphate ban legislation in many countries during the 1970s-1980s led to reformulation using alternative builders including **zeolites**, **polycarboxylates**, and **silicates**. **Sodium aluminum silicate (zeolite A)** provides ion-exchange capacity for hardness removal:

$$Na_{12}Al_{12}Si_{12}O_{48} + 6Ca^{2+} \rightarrow Ca_6Al_{12}Si_{12}O_{48} + 12Na^+ \quad (2.115)$$

Nitrogen compounds in detergent formulations can also contribute to eutrophication, particularly in marine systems where nitrogen often limits primary productivity. **Quaternary ammonium compounds** used as fabric softeners and disinfectants undergo biodegradation to release ammonia and nitrate that support algal growth.

Surfactant effects on eutrophication involve both direct toxicity to aquatic organisms and indirect effects on nutrient cycling. **Linear alkylbenzene sulfonates (LAS)** and **alcohol ethoxylates (AE)** exhibit varying biodegradation rates and aquatic toxicity depending on their chemical structure and environmental conditions.

Foam formation from surfactants can reduce atmospheric oxygen transfer rates and create anaesthetic conditions that promote nutrient release from sediments. **Protein denaturation** by surfactants affects aquatic organisms and microbial communities involved in nutrient cycling.

Algal growth kinetics follow Monod relationships for nutrient-limited conditions:

$$\mu = \mu\_max \times [P]/([P] + K\_s) \quad (2.116)$$

where μ is specific growth rate, μ_max is maximum growth rate, [P] is phosphorus concentration, and K_s is the half-saturation constant. **Liebig's Law of the Minimum** determines which nutrient limits growth in multi-nutrient systems.

Redfield ratios describe the stoichiometric relationships between carbon, nitrogen, and phosphorus in algal biomass (C:N:P = 106:16:1 by atoms). **Nutrient limitation patterns** vary between freshwater systems (typically phosphorus-limited) and marine systems (typically nitrogen-limited), affecting the relative importance of different detergent constituents.

Harmful algal blooms (HABs) represent extreme manifestations of eutrophication, involving species such as *Microcystis*, *Anabaena*, and *Alexandrium* that produce potent toxins. **Cyanotoxins** including microcystin, anatoxin, and cylindrospermopsin pose serious health risks through drinking water contamination and recreational water contact.

Hypoxia development follows eutrophication through the decomposition of algal biomass:

$C_{106}H_{263}O_{110}N_{16}P + 138O_2 \rightarrow 106CO_2 + 122H_2O + 16HNO_3 + H_3PO_4$ (2.117)

Oxygen depletion creates conditions favoring anaerobic bacteria and the release of reduced compounds including hydrogen sulfide, methane, and phosphine that further degrade water quality.

Lake recovery following nutrient loading reductions demonstrates the persistence of eutrophication effects. **Internal phosphorus loading** from sediments can sustain eutrophic conditions for decades after external loading reductions, particularly in lakes with long hydraulic retention times and large sediment phosphorus pools.

Sediment phosphorus release occurs under anaerobic conditions through the reduction of iron-bound phosphates:

$Fe^{3+}\text{-}PO_4 + \text{organic matter} \rightarrow Fe^{2+} + PO_4^{3-} + CO_2$ (2.118)

Lake restoration techniques including aluminumsulfate treatment, sediment dredging, and hypolimnetic aeration address internal phosphorus loading but require long-term commitment and substantial financial resources.

Marine eutrophication exhibits different characteristics from freshwater systems, with nitrogen limitation promoting different algal species and creating extensive coastal dead zones. **Gulf of Mexico hypoxia** represents one of the largest dead zones globally, linked to agricultural nutrient runoff from the Mississippi River watershed.

Ecosystem services degraded by eutrophication include water supply, fisheries, recreation, and biodiversity conservation. **Economic costs** of eutrophication in the United States exceed $2 billion annually through water treatment, recreation losses, and ecosystem restoration efforts.

Alternative detergent technologies continue to evolve to minimize environmental impacts. **Enzyme-based formulations** reduce the need for high-temperature washing and harsh chemical builders. **Concentrated formulations**reduce packaging materials and transportation impacts. **Plant-based surfactants** from renewable feedstocks offer improved biodegradability compared to petroleum-derived alternatives.

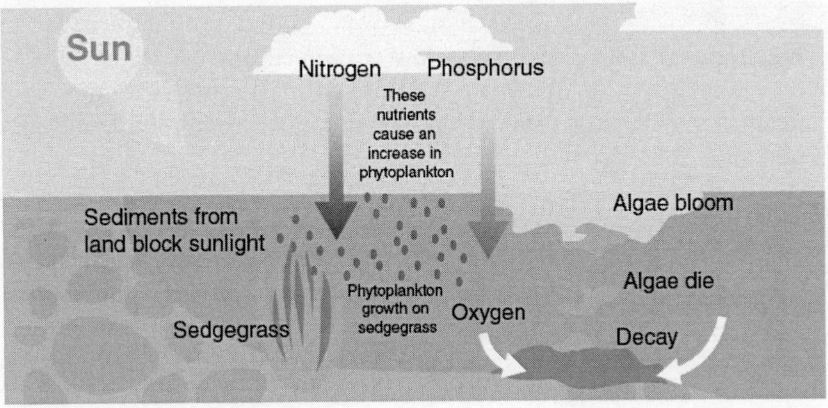

Fig. 11: Conceptual diagram showing the relationship between detergent phosphorus inputs, eutrophication processes, and ecosystem impacts in freshwater systems

2.5 Soil Chemistry and Fertilizer Science

Soil Composition and Nutrient Cycles

Soil represents a complex, dynamic system that serves as the foundation for terrestrial ecosystems and agricultural productivity. **Soil composition** involves four major components: mineral particles (45% by volume), organic matter (5%), water (25%), and air (25%), though these proportions vary significantly with soil type, management practices, and environmental conditions. Understanding soil chemistry is essential for sustainable agriculture, environmental protection, and ecosystem management.

Mineral components originate from the weathering of parent rocks and determine many fundamental soil properties. **Primary minerals** including quartz, feldspars, and micas reflect the composition of source rocks and exhibit varying resistance to weathering processes. **Quartz (SiO_2)** demonstrates exceptional stability and dominates the sand fraction in many soils, while **feldspars** weather more readily to release potassium, sodium, and calcium ions.

Secondary minerals form through weathering processes and exhibit much higher chemical reactivity than primary minerals. **Clay minerals** including kaolinite, montmorillonite, and illite possess layered crystal structures with high surface areas and cation exchange capacities. **Kaolinite** exhibits relatively low cation exchange capacity (3-15 cmol/kg) due to its 1:1 layer structure, while **montmorillonite** shows much higher values (80-150 cmol/kg) due to isomorphic substitution within its 2:1 layer structure.

Soil organic matter (SOM) comprises living biomass, fresh plant and animal residues, and stabilized humic substances formed through decomposition processes. **Humic substances** include humic acids, fulvic acids, and humin, characterized by complex aromatic and aliphatic structures with numerous functional groups capable of metal complexation and nutrient retention.

Cation exchange capacity (CEC) represents one of the most important soil chemical properties, quantifying the soil's ability to retain and exchange positively charged nutrients. **Exchange reactions** follow mass action principles:

$$\text{Clay-Ca} + 2K^+ \leftrightarrow \text{Clay-2K} + Ca^{2+} \quad (2.119)$$

Selectivity coefficients describe the relative affinity of different cations for exchange sites, generally following the order: $Al^{3+} > Ca^{2+} > Mg^{2+} > K^+ > NH_4^+ > Na^+$ for most soil colloids.

Soil pH influences virtually all soil chemical and biological processes through its effects on mineral solubility, microbial activity, and nutrient availability. **Buffer capacity** describes the soil's resistance to pH change and depends on organic matter content, clay mineralogy, and carbonate presence. **Lime requirement** calculations utilize buffer pH measurements to determine the amount of agricultural limestone needed to achieve target pH levels.

Aluminum toxicity becomes problematic in acid soils ($pH < 5.5$) where Al^{3+} solubility increases dramatically. **Hydrolysis reactions** of aluminum generate additional acidity:

$$Al^{3+} + H_2O \rightarrow Al(OH)^{2+} + H^+ \quad (2.120)$$

Liming materials including calcium carbonate and magnesium carbonate neutralize soil acidity and precipitate toxic aluminum:

$$CaCO_3 + 2H^+ \rightarrow Ca^{2+} + H_2O + CO_2 \qquad (2.121)$$

Nutrient cycling in soils involves complex interactions between chemical, biological, and physical processes that regulate nutrient availability for plant uptake. **Nitrogen cycling** represents one of the most dynamic and environmentally significant processes, involving multiple chemical transformations mediated by diverse microbial communities.

Nitrogen mineralization converts organic nitrogen to plant-available forms through decomposition of proteins, nucleic acids, and other nitrogen-containing compounds. **Ammonification** represents the initial step where organic nitrogen is converted to ammonium:

$$R\text{-}NH_2 + H_2O \rightarrow NH_4^+ + R\text{-}OH + OH^- \qquad (2.122)$$

Nitrification follows through the sequential oxidation of ammonium to nitrate by specialized autotrophic bacteria. **Ammonia-oxidizing bacteria (AOB)** including *Nitrosomonas* species catalyze the first step:

$$NH_4^+ + 1.5O_2 \rightarrow NO_2^- + 2H^+ + H_2O \qquad (2.123)$$

Nitrite-oxidizing bacteria (NOB) such as *Nitrobacter* complete the process:

$$NO_2^- + 0.5O_2 \rightarrow NO_3^- \qquad (2.124)$$

Denitrification occurs under anaerobic conditions when bacteria use nitrate as an electron acceptor for respiration, potentially leading to nitrogen losses through gaseous emissions:

$$2NO_3^- + 10H^+ + 10e^- \rightarrow N_2 + 5H_2O \qquad (2.125)$$

Phosphorus cycling involves transformations between organic and inorganic forms, with **phosphatase enzymes** playing crucial roles in organic phosphorus mineralization. **Inorganic phosphorus** exists in various chemical forms depending on soil pH and mineral composition. **Iron and aluminum phosphates** dominate in acid soils:

$$Fe^{3+} + PO_4^{3-} \rightarrow FePO_4 \quad (2.126)$$

$$Al^{3+} + PO_4^{3-} \rightarrow AlPO_4 \quad (2.127)$$

Calcium phosphates become increasingly important at higher pH values:

$$3Ca^{2+} + 2PO_4^{3-} \rightarrow Ca_3(PO_4)_2 \quad (2.128)$$

Phosphorus sorption by soil colloids represents a major factor controlling phosphorus availability, with **maximum sorption capacity** related to iron and aluminum oxide content. **Langmuir isotherms** describe phosphorus sorption behavior:

$$q = (Q\_max \times K\_L \times C)/(1 + K\_L \times C) \quad (2.129)$$

where q is sorbed phosphorus, Q_max is maximum sorption capacity, K_L is the Langmuir constant, and C is solution phosphorus concentration.

Sulfur cycling involves oxidation-reduction reactions that affect sulfur availability and soil chemistry. **Sulfatereduction**under anaerobic conditions produces hydrogen sulfide:

$$SO_4^{2-} + 8H^+ + 8e^- \rightarrow S^{2-} + 4H_2O \quad (2.130)$$

Sulfur oxidation by *Thiobacillus* species converts reduced sulfur compounds back to sulfate while generating acidity:

$$S^0 + 1.5O_2 + H_2O \rightarrow SO_4^{2-} + 2H^+ \quad (2.131)$$

Potassium cycling involves relatively simple chemical processes due to the stability of the K^+ ion. **Potassium fixation**occurs through entrapment in clay mineral interlayers, particularly in illite and vermiculite:

$$K^+ + Clay^- \rightarrow Clay\text{-}K\ (fixed) \quad (2.132)$$

Micronutrient cycling involves complex interactions with organic matter and pH-dependent solubility changes. **Iron availability** decreases dramatically with increasing pH due to precipitation of iron hydroxides:

$Fe^{3+} + 3OH^- \rightarrow Fe(OH)_3$ (2.133)

Chelation by organic compounds enhances micronutrient availability through the formation of stable, soluble complexes that prevent precipitation and facilitate plant uptake.

Classification and Application of Fertilizers (Urea, DAP)

Fertilizer science represents the application of chemistry to enhance plant nutrition and agricultural productivity through the strategic addition of essential nutrients to soil systems. **Fertilizer classification** follows several schemes based on nutrient content, chemical composition, physical form, and release characteristics. Understanding these classifications is essential for optimizing fertilizer use efficiency and minimizing environmental impacts.

Macronutrient fertilizers supply nitrogen, phosphorus, and potassium in various chemical forms and combinations. **Nitrogen fertilizers** represent the largest category by volume and include ammonium-based, nitrate-based, and urea-based materials. **Phosphorus fertilizers** primarily consist of processed phosphate rock products, while **potassium fertilizers** are derived from underground deposits of potassium salts.

Urea ($(NH_2)_2CO$) represents the most widely used nitrogen fertilizer globally, accounting for over 50% of world nitrogen fertilizer consumption. **Urea synthesis** follows the Haber-Bosch process for ammonia production followed by reaction with carbon dioxide:

$2NH_3 + CO_2 \rightarrow (NH_2)_2CO + H_2O$ (2.134)

Physical properties of urea include high nitrogen content (46% N), excellent water solubility (1080 g/L at 20°C), and low salt index, making it suitable for diverse application methods and crop types. **Hygroscopic behavior** requires careful storage to prevent moisture absorption and caking.

Urea hydrolysis in soil occurs through the enzyme urease, which is widespread in soil microorganisms and plant residues:

$(NH_2)_2CO + H_2O \rightarrow 2NH_3 + CO_2$ (2.135)

Ammonia volatilization represents a major loss mechanism when urea is surface-applied without incorporation, particularly under high pH and temperature conditions. **Volatilization losses** can exceed 30% of applied nitrogen under adverse conditions, reducing fertilizer efficiency and contributing to atmospheric ammonia pollution.

Urease inhibitors including N-(n-butyl) thiophosphorictriamide (NBPT) can reduce ammonia volatilization by temporarily blocking urease activity:

$$\text{Urease} + \text{NBPT} \rightarrow \text{Urease-NBPT (inactive complex)} \quad (2.136)$$

Nitrification inhibitors such as dicyandiamide (DCD) and 3,4-dimethylpyrazole phosphate (DMPP) slow the conversion of ammonium to nitrate, reducing leaching losses and improving nitrogen use efficiency.

Diammonium phosphate (DAP) serves as a major phosphorus fertilizer, containing 18% nitrogen and 46% P_2O_5 (20% P). **DAP production** involves the reaction of ammonia with phosphoric acid:

$$2NH_3 + H_3PO_4 \rightarrow (NH_4)_2HPO_4 \quad (2.137)$$

Soil reactions of DAP involve dissolution and subsequent interactions with soil components. **Initial dissolution** creates a zone of high pH around the fertilizer granule:

$$(NH_4)_2HPO_4 \rightarrow 2NH_4^+ + HPO_4^{2-} \quad (2.138)$$

Ammonium nitrification generates acidity that can lower soil pH in the fertilizer reaction zone:

$$NH_4^+ + 2O_2 \rightarrow NO_3^- + 2H^+ + H_2O \quad (2.139)$$

Phosphorus reactions in soil include precipitation with iron, aluminum, and calcium depending on soil pH conditions. **Phosphorus fixation** reduces the availability of applied phosphorus over time, particularly in highly weathered soils with high iron and aluminum oxide content.

Monoammonium phosphate (MAP) provides an alternative phosphorus source with 11% nitrogen and 52% P_2O_5 (23% P). **MAP**

characteristics include higher phosphorus content than DAP but lower pH of the fertilizer solution, which can be advantageous in high-pH soils.

Controlled-release fertilizers employ various technologies to regulate nutrient release rates and improve synchronization with plant nutrient demand. **Polymer-coated fertilizers** encapsulate conventional fertilizers with semi-permeable membranes that control water penetration and nutrient diffusion.

Release mechanisms for polymer-coated fertilizers involve **osmotic pressure** driving water into the coating, dissolution of the fertilizer core, and diffusion of nutrients through the coating:

$$\text{Rate} \propto \Delta C \times D \times A/\delta \quad (2.140)$$

where ΔC is the concentration gradient, D is the diffusion coefficient, A is the surface area, and δ is the coating thickness.

Slow-release fertilizers including urea-formaldehyde and isobutylidenediurea (IBDU) undergo gradual hydrolysis to release nitrogen over extended periods. **Urea-formaldehyde** synthesis creates methylene urea polymers with varying chain lengths:

$$nCO(NH_2)_2 + nCH_2O \rightarrow -(CONH\text{-}CH_2\text{-}NH)_n\text{-} + nH_2O \quad (2.141)$$

Fertilizer placement strategies significantly affect nutrient use efficiency and environmental impacts. **Band placement**concentrates fertilizer in narrow zones near seed or plant roots, reducing soil contact and fixation reactions while improving plant access to nutrients.

Split applications divide total fertilizer requirements into multiple applications timed to match plant nutrient demand patterns. **Variable rate technology** utilizes GPS and soil testing data to apply fertilizers at site-specific rates within fields, optimizing nutrient management and reducing environmental losses.

Precision agriculture integrates multiple technologies including soil testing, plant tissue analysis, remote sensing, and GPS-guided application equipment to optimize fertilizer use efficiency. **Yield goal approaches** calculate fertilizer requirements based on realistic yield targets and nutrient removal rates:

Fertilizer need = (Yield goal × Nutrient removal) - Soil supply (2.142)

Soil Macronutrients and Micronutrients

Plant nutrition requires seventeen essential elements that must be available in adequate quantities and appropriate chemical forms for optimal growth and development. **Macronutrients** including nitrogen, phosphorus, potassium, calcium, magnesium, and sulfur are required in relatively large quantities, while **micronutrients** including iron, manganese, zinc, copper, boron, molybdenum, and chlorine are needed in much smaller amounts but are equally essential for plant function.

Nitrogen functions in plants include protein synthesis, chlorophyll formation, and nucleic acid production. **Nitrogen deficiency symptoms** typically appear as yellowing of older leaves (chlorosis) due to the mobility of nitrogen within plants and its redistribution from older to younger tissues. **Excessive nitrogen** can delay maturity, increase disease susceptibility, and reduce root development.

Nitrogen forms available for plant uptake include ammonium (NH_4^+), nitrate (NO_3^-), and to a lesser extent, organic nitrogen compounds such as amino acids. **Ammonium uptake** is metabolically less expensive than nitrate uptake because it doesn't require reduction before incorporation into amino acids:

$$NH_4^+ + \alpha\text{-ketoglutarate} + NADH \rightarrow \text{glutamate} + NAD^+ \qquad (2.143)$$

Nitrate reduction occurs through the nitrate reductase and nitrite reductase enzyme systems:

$$NO_3^- + NADH + H^+ \rightarrow NO_2^- + NAD^+ + H_2O \qquad (2.144)$$

$$NO_2^- + 6\ \text{ferredoxin(red)} + 8H^+ \rightarrow NH_4^+ + 6\ \text{ferredoxin(ox)} + 2H_2O \qquad (2.145)$$

Phosphorus functions include energy transfer through ATP and ADP, genetic material composition (DNA and RNA), and membrane structure through phospholipids. **Phosphorus deficiency** causes stunted growth, delayed maturity, and purpling of older leaves in some species. **Root:shoot ratios** typically increase under phosphorus deficiency as plants allocate more resources to root growth for enhanced nutrient acquisition.

Phosphorus chemistry in plants involves numerous organophosphate compounds including ATP, ADP, sugar phosphates, and phosphorylated proteins. **Phosphorus remobilization** from older plant tissues allows continued growth under phosphorus-limiting conditions, though this redistribution is less complete than for nitrogen.

Potassium functions include enzyme activation, osmoregulation, and stomatal control. **Potassium deficiency** symptoms include marginal leaf burn (scorch) on older leaves and increased susceptibility to drought stress and diseases. **Potassium requirements** are particularly high for crops that produce large amounts of carbohydrates, such as potatoes and sugar beets.

Potassium transport in plants occurs primarily through potassium channels and transporters that maintain cellular K^+ concentrations around 100-150 mM.**Osmotic adjustment** involves potassium accumulation in response to water stress, helping maintain cell turgor and physiological function.

Calcium functions include cell wall formation, membrane stability, and enzyme regulation. **Calcium deficiency**disorders include blossom end rot in tomatoes, bitter pit in apples, and tip burn in lettuce. **Calcium mobility** within plants is extremely limited due to its structural role in cell walls and its precipitation with organic acids.

Calcium signaling involves rapid changes in cytoplasmic Ca^{2+} concentrations that regulate numerous physiological processes including responses to environmental stress and pathogen attack. **Calmodulin** and other calcium-binding proteins mediate these cellular responses.

Magnesium serves as the central atom in chlorophyll molecules and activates numerous enzymes involved in phosphate metabolism. **Magnesium deficiency** causes interveinal chlorosis of older leaves, with the characteristic yellowing pattern reflecting the distribution of chlorophyll-containing tissues.

Magnesium-ATP complexes represent the active form of ATP for most enzymatic reactions, highlighting the importance of adequate magnesium nutrition for energy metabolism. **Magnesium transport** in soils and plants involves competition with other divalent cations, particularly calcium.

Sulfur functions include protein synthesis through sulfur-containing amino acids (cysteine and methionine) and the formation of disulfide bonds that stabilize protein structure. **Sulfur deficiency** symptoms resemble nitrogen deficiency but typically appear first on younger leaves due to sulfur's limited mobility within plants.

Iron functions as a component of numerous enzymes and electron transport proteins, particularly in photosynthesis and respiration. **Iron deficiency chlorosis** represents a widespread nutritional disorder, especially in high-pH soils where iron availability is limited by precipitation reactions.

Iron chelation by organic compounds including citrate, malate, and specialized siderophores enhances iron solubility and plant availability:

$Fe^{3+} + citrate^{3-} \rightarrow [Fe\text{-}citrate]$ (2.146)

Strategy I plants (dicots and non-grass monocots) respond to iron deficiency by acidifying the rhizosphere, reducing ferric iron to ferrous iron, and increasing iron uptake capacity. **Strategy II plants** (grasses) secrete phytosiderophores that chelate iron and facilitate its uptake through specific transporters.

Zinc functions include enzyme activation, protein synthesis, and membrane stability. **Zinc deficiency** symptoms include stunted growth, shortened internodes, and chlorotic mottling of younger leaves. **Zinc availability** decreases with increasing soil pH due to precipitation and adsorption reactions.

Manganese activates enzymes involved in photosystem II and various metabolic pathways. **Manganese deficiency**causes interveinal chlorosis similar to iron deficiency but typically affects middle-aged leaves first. **Manganese toxicity**can occur in acid soils where Mn^{2+} solubility is high.

Copper functions include electron transport, lignin synthesis, and superoxide dismutase activity. **Copper deficiency** is relatively rare but can cause wilting, chlorosis, and die-back of shoot tips. **Copper availability** is strongly influenced by organic matter content due to its tendency to form stable organic complexes.

Boron functions include cell wall formation, membrane function, and pollen viability. **Boron deficiency** causes brittle stems, heart rot in root crops, and poor fruit set. **Boron mobility** within plants is species-dependent, with limited mobility in most crops leading to deficiency symptoms in growing points.

Molybdenum serves as a component of nitrogenase and nitrate reductase enzymes. **Molybdenum deficiency** primarily affects nitrogen metabolism, particularly in legumes where biological nitrogen fixation is impaired. **Molybdenum availability** increases with increasing soil pH, opposite to the behavior of most other micronutrients.

Pollution Due to Pesticides and Heavy Metals

Soil contamination by pesticides and heavy metals represents a major environmental concern with implications for food safety, ecosystem health, and groundwater quality. **Pesticide pollution** results from both intentional applications for pest control and accidental spills or improper disposal practices. **Heavy metal contamination** originates from industrial activities, mining operations, sewage sludge application, and atmospheric deposition.

Pesticide behavior in soils involves complex interactions between chemical properties, soil characteristics, and environmental conditions. **Sorption processes** determine the distribution of pesticides between soil solid and liquid phases, affecting their mobility, bioavailability, and persistence. **Organic matter content** represents the primary factor controlling pesticide sorption for most compounds:

$$K_d = K_{oc} \times f_{oc} \quad (2.147)$$

where K_d is the soil-water distribution coefficient, K_{oc} is the organic carbon partition coefficient, and f_{oc} is the fraction of organic carbon.

Pesticide degradation occurs through biotic and abiotic processes that determine environmental persistence and accumulation potential. **Hydrolysis reactions** represent important abiotic degradation pathways for many pesticides, with reaction rates strongly influenced by pH and temperature:

$$\text{Rate} = k \times [\text{pesticide}] \times [H^+]^n \quad (2.148)$$

Microbial degradation typically represents the primary degradation pathway for most organic pesticides. **Aerobic degradation** generally proceeds more rapidly than anaerobic degradation, though some compounds show enhanced degradation under reducing conditions.

Organochlorine pesticides including DDT, chlordane, and dieldrin exhibit exceptional persistence and bioaccumulation potential due to their chemical stability and lipophilic properties. **DDT degradation** proceeds slowly through dehydrochlorination to DDE:

$$\text{DDT} \rightarrow \text{DDE} + \text{HCl} \quad (2.149)$$

Bioaccumulation factors for organochlorine pesticides can exceed 1000-fold concentration increases through food chains, leading to toxic effects in higher trophic levels including reproductive failure in birds and mammals.

Organophosphate pesticides generally show lower environmental persistence but exhibit high acute toxicity through acetylcholinesterase inhibition. **Hydrolysis half-lives** typically range from days to weeks depending on pH and temperature conditions. **Metabolites** from organophosphate degradation can retain significant toxicity.

Carbamate pesticides share the same mode of action as organophosphates but generally exhibit greater chemical instability and shorter environmental persistence. **N-methylcarbamates** undergo hydrolysis and photodegradation relatively rapidly in most environmental conditions.

Triazine herbicides including atrazine and simazine exhibit moderate persistence and have been detected in groundwater supplies worldwide. **Atrazine degradation** occurs primarily through dealkylation and hydroxylation reactions mediated by soil microorganisms:

$$\text{Atrazine} \rightarrow \text{Desethylatrazine} \rightarrow \text{Hydroxyatrazine} \quad (2.150)$$

Heavy metal pollution in soils involves elements including lead, cadmium, mercury, chromium, copper, zinc, and nickel at concentrations that exceed natural background levels and pose risks to human health or ecosystem function. **Heavy metal sources** include industrial emissions, mining activities, sewage sludge application, fertilizer impurities, and atmospheric deposition from combustion processes.

Lead contamination in urban soils often originates from historical use of leaded gasoline and lead-based paints. **Lead bioavailability** depends on soil pH, organic matter content, and the presence of phosphate minerals that can immobilize lead through precipitation reactions:

$$3Pb^{2+} + 2PO_4^{3-} \rightarrow Pb_3(PO_4)_2 \qquad (2.151)$$

Cadmium pollution results primarily from industrial activities, phosphate fertilizers, and sewage sludge application. **Cadmium mobility** in soils increases with decreasing pH and low organic matter content. **Plant uptake** of cadmium occurs readily through divalent cation transporters, leading to accumulation in edible plant tissues.

Mercury contamination involves both elemental mercury and organomercury compounds, with methylmercury representing the most toxic and bioaccumulative form. **Mercury methylation** occurs under anaerobic conditions through bacterial processes:

$$Hg^{2+} + CH_3^- \rightarrow CH_3Hg^+ \qquad (2.152)$$

Chromium contamination involves two primary oxidation states with dramatically different toxicological properties. **Hexavalent chromium (Cr^{6+})** exhibits high solubility, mobility, and toxicity, while **trivalent chromium (Cr^{3+})** is relatively immobile and less toxic. **Chromium reduction** can occur through both chemical and biological pathways:

$$Cr_2O_7^{2-} + 14H^+ + 6e^- \rightarrow 2Cr^{3+} + 7H_2O \qquad (2.153)$$

Soil remediation technologies for heavy metal contamination include physical removal, chemical immobilization, and biological treatment approaches. **Soil washing** employs chemical extractants to solubilize and remove metals from contaminated soils. **In-situ stabilization** uses amendments such as lime, phosphate, and organic matter to reduce metal bioavailability:

$$Pb^{2+} + HPO_4^{2-} \rightarrow PbHPO_4 \qquad (2.154)$$

Phytoremediation utilizes plants to extract, stabilize, or degrade soil contaminants. **Hyperaccumulator plants** can accumulate extraordinarily high concentrations of specific metals in their tissues. **Phytoextraction** removes metals from soil through plant uptake and harvesting, while **phytostabilization** reduces metal mobility through root exudates and rhizosphere processes.

Risk assessment for soil contamination considers exposure pathways including direct soil ingestion, inhalation of contaminated dust, and consumption of contaminated food crops. **Bioavailability** represents a critical factor in risk assessment, as total contaminant concentrations may overestimate actual health risks.

Regulatory frameworks for soil contamination vary widely among countries but generally establish **screening levels**based on land use scenarios and **cleanup standards** for contaminated sites. **Background concentrations** provide reference points for distinguishing between natural and anthropogenic contamination sources.

Key Takeaways

Environmental chemistry serves as a critical bridge between fundamental chemical principles and real-world environmental challenges. **Air pollution control** requires understanding of complex atmospheric chemistry involving primary and secondary pollutant formation, with catalytic converters representing successful applications of chemical principles to emission reduction. **Climate change mitigation** depends on comprehensive knowledge of greenhouse gas chemistry and atmospheric processes.

- **Water quality assessment** integrates analytical chemistry with public health protection, utilizing techniques from classical titrations to advanced instrumental methods. **BOD and COD measurements** provide complementary information about organic pollution, while **nutrient analysis** guides eutrophication control strategies. **Water treatment technologies** demonstrate practical applications of chemical precipitation, biological processes, and advanced oxidation.

- **Soil chemistry fundamentals** govern nutrient cycling, fertilizer effectiveness, and contaminant fate in terrestrial systems. **Sustainable fertilizer use** requires balancing crop nutrition needs with environmental protection through precision application technologies and enhanced-efficiency fertilizers. **Soil**

contamination by pesticides and heavy metals poses long-term challenges requiring integrated remediation approaches.

The future of environmental chemistry lies in developing **green technologies** that prevent pollution at its source rather than treating it after formation. **Circular economy principles** emphasize resource recovery and waste minimization through chemical process innovations. **Climate adaptation strategies** will increasingly rely on environmental chemistry applications to address emerging challenges including water scarcity, extreme weather events, and ecosystem shifts.

Understanding environmental chemistry is essential for **sustainable development** that meets human needs while protecting environmental quality for future generations. The integration of chemical knowledge with environmental science, engineering, and policy provides the foundation for addressing complex environmental challenges at local, regional, and global scales.

CHAPTER 3
BIOCHEMICAL COMPOUNDS AND MEDICINAL CHEMISTRY

Learning Objectives

Upon completion of this chapter, students will be able to: Analyze the structural diversity and biological functions of carbohydrates, including monosaccharides and disaccharides • Evaluate the chemical stability and nutritional significance of vitamins in food systems and human health • Understand the structure-activity relationships in pharmaceutical compounds and their mechanisms of action • Assess the chemistry of antibiotics and antiparasitic agents in disease control • Examine the chemical properties of lipids, oils, and surfactants in biological and industrial applications

Introduction

The intersection of chemistry and biology forms the foundation of modern biochemistry and medicinal chemistry, disciplines that have revolutionized our understanding of life processes and therapeutic interventions. **Biochemical compounds** encompass a vast array of molecules that participate in cellular processes, from simple sugars that provide energy to complex pharmaceutical agents that modulate disease pathways. This chapter explores the intricate chemical world of biomolecules, examining their structures, functions, and practical applications in nutrition, healthcare, and industrial processes.

The study of biochemical compounds represents one of the most dynamic areas in modern chemistry, where molecular structure directly correlates with biological function. **Medicinal chemistry**, as a discipline, emerged in the mid-20th century through the pioneering work of scientists like Sir James Black, who developed beta-blockers and H_2-receptor antagonists, fundamentally changing cardiovascular and gastric disease treatment. The systematic approach to drug design has evolved from empirical observations to sophisticated computational modeling and structure-based design principles.

Contemporary biochemical research integrates multiple analytical techniques, from traditional polarimetry for sugar analysis to advanced nuclear magnetic resonance (NMR) spectroscopy for elucidating complex molecular structures. The pharmaceutical industry's evolution from natural product isolation to synthetic drug development reflects humanity's growing understanding of molecular interactions and their therapeutic potential. This chapter provides a comprehensive examination of these fundamental biochemical systems, emphasizing their practical applications in food science, environmental monitoring, and healthcare delivery.

3.1 Carbohydrate Chemistry

Fundamental Structure and Classification

Carbohydrates constitute the most abundant class of organic compounds in nature, serving as primary energy sources and structural components in biological systems. The fundamental chemistry of carbohydrates centers on their polyhydroxy aldehyde or ketone nature, with the general formula $C_n(H_2O)_n$, although this simplification does not capture the structural diversity present in this molecular class.

Monosaccharides represent the simplest carbohydrate units and serve as building blocks for more complex structures. **Glucose** ($C_6H_{12}O_6$), the most physiologically important monosaccharide, exists predominantly in the D-configuration in biological systems. The molecule adopts a pyranose ring structure through intramolecular hemiacetal formation between the aldehyde group at C-1 and the hydroxyl group at C-5. This cyclization creates two distinct anomeric forms: α-D-glucopyranose and β-D-glucopyranose, which differ in the spatial orientation of the hydroxyl group at the anomeric carbon.

The **α-anomer** exhibits the hydroxyl group in the axial position relative to the ring plane, while the **β-anomer** positions this group equatorially. This stereochemical difference profoundly influences the biological properties and intermolecular interactions of glucose derivatives. In aqueous solutions, these anomers exist in dynamic equilibrium through the linear aldehyde form, a phenomenon known as **mutarotation**.

Fructose, the sweetest naturally occurring sugar, adopts a furanose ring structure predominantly, forming β-D-fructofuranose through hemiacetal formation between the ketone group at C-2 and the hydroxyl group at C-5. The unique structural features of fructose contribute to its distinct metabolic pathway and its role in modern concerns about high-fructose corn syrup consumption.

Disaccharide Chemistry and Glycosidic Bonds

Disaccharides form through glycosidic bond formation between two monosaccharide units, creating molecules with diverse biological functions and industrial applications. **Maltose**, composed of two glucose units linked by an α(1→4) glycosidic bond, represents the primary product of starch hydrolysis by α-amylase enzymes. The structural characteristics of maltose allow for further enzymatic cleavage by maltase, making it an important intermediate in carbohydrate metabolism.

Lactose, the principal sugar in mammalian milk, consists of galactose and glucose linked by a β(1→4) glycosidic bond. This disaccharide presents unique challenges in food chemistry due to **lactose intolerance**, a condition affecting a significant portion of the global population due to decreased lactase enzyme activity in adulthood. The chemistry of lactose hydrolysis has led to the development of lactase-treated dairy products and alternative milk formulations.

The glycosidic bond formation can be represented by the general condensation reaction:

$$\text{Monosaccharide}_1 + \text{Monosaccharide}_2 \rightarrow \text{Disaccharide} + H_2O \quad (3.1)$$

Mutarotation Phenomena and Kinetics

Mutarotation represents a fundamental property of reducing sugars, involving the interconversion between anomeric forms through the open-chain intermediate. This process occurs spontaneously in aqueous solutions and follows first-order kinetics. The equilibrium composition for glucose in water at 25°C consists of approximately 36% α-anomer and 64% β-anomer, with less than 1% existing in the open-chain form.

The kinetics of mutarotation can be monitored through **polarimetry**, as each anomeric form exhibits distinct optical rotation values. The α-anomer of glucose shows a specific rotation of +112°, while the β-anomer exhibits +18.7°. The equilibrium specific rotation reaches +52.5°, representing the weighted average of the anomeric mixture.

The rate of mutarotation depends on several factors including pH, temperature, and the presence of catalysts. **Acid and base catalysis** significantly accelerate the process, with optimal rates occurring at pH values around 4-5. This pH dependence reflects the mechanism involving protonation of the ring oxygen (acid catalysis) or deprotonation of hydroxyl groups (base catalysis), facilitating ring opening and subsequent reclosure.

Reducing Sugar Tests and Analytical Methods

The analytical chemistry of carbohydrates relies heavily on the **reducing sugar** concept, which describes sugars capable of reducing metal ions or other oxidizing agents due to their free anomeric carbon. Classical reducing sugar tests include the **Benedict's test** and **Fehling's test**, both based on the reduction of Cu^{2+} ions to Cu_2O under alkaline conditions.

The Benedict's reagent contains copper sulfate, sodium citrate, and sodium carbonate in aqueous solution. Upon heating with reducing sugars, the blue Cu^{2+} ions are reduced to red-orange Cu_2O precipitate:

$$RCHO + 2Cu^{2+} + 5OH^- \rightarrow RCOO^- + Cu_2O + 3H_2O \qquad (3.2)$$

Modern analytical techniques for carbohydrate analysis include high-performance liquid chromatography (HPLC) with refractive index detection, gas chromatography-mass spectrometry (GC-MS) of derivatized samples, and enzymatic assays specific for individual sugars. These methods provide quantitative analysis with high specificity and sensitivity, essential for food quality control and clinical diagnostics.

Polarimetry and NMR Applications

Polarimetry serves as a classical yet still valuable technique for carbohydrate analysis, exploiting the optical activity inherent in these chiral molecules. The specific rotation $[\alpha]$ is defined by:

$$[\alpha] = \alpha / (l \times c) \qquad (3.3)$$

where α is the observed rotation, l is the path length in decimeters, and c is the concentration in g/mL.

Nuclear Magnetic Resonance (NMR) spectroscopy has revolutionized carbohydrate structural analysis, providing detailed information about ring conformation, anomeric configuration, and glycosidic linkage patterns. ^{13}C NMR spectroscopy proves particularly valuable, as the anomeric carbon signals appear in distinct regions: α-anomers typically resonate around 93-102 ppm, while β-anomers appear at 97-105 ppm.

Two-dimensional NMR techniques, including COSY (Correlation Spectroscopy) and TOCSY (Total Correlation Spectroscopy), enable complete assignment of complex carbohydrate structures. These advanced methods have been instrumental in elucidating the structures of oligosaccharides and polysaccharides in biological systems.

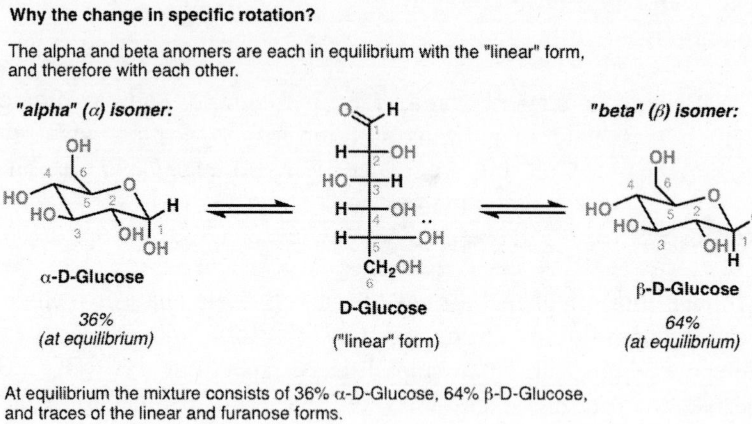

Fig. 12: Mutarotation equilibrium of glucose showing α and β anomeric forms with their respective optical rotation values and equilibrium percentages

3.2 Vitamins: Structure and Deficiency

Classification and Solubility Properties

Vitamins represent a diverse group of organic compounds essential for normal physiological function, required in small quantities, and generally not synthesized by the human body in sufficient amounts. The classification of vitamins into **water-soluble** and **fat-soluble** categories reflects their fundamental chemical properties and has profound implications for their absorption, transport, storage, and excretion.

Water-soluble vitamins include the B-complex group and vitamin C (ascorbic acid). These compounds contain polar functional groups such as hydroxyl, amino, and carboxyl groups that facilitate hydrogen bonding with water molecules. Their hydrophilic nature results in rapid absorption from the gastrointestinal tract, efficient renal excretion, and minimal body storage capacity. Consequently, regular dietary intake becomes crucial to prevent deficiency states.

Fat-soluble vitamins (A, D, E, and K) contain predominantly nonpolar hydrocarbon structures with limited polar functional groups. These vitamins require lipid-containing vehicles for absorption and transport, utilizing the same mechanisms as dietary fats. They are stored in adipose tissue and the liver, providing reserves that can sustain the body during periods of inadequate intake. However, this storage capacity also increases the risk of toxicity with excessive consumption.

Vitamin A: Retinoid Chemistry and Visual Function

Vitamin A encompasses a group of compounds collectively known as **retinoids**, including retinol, retinal, and retinoic acid. The fundamental structure consists of a β-ionone ring connected to an isoprenoid chain containing alternating double bonds in a trans-configuration. **Retinol** (vitamin A_1) serves as the primary storage form, while **retinal** functions as the active chromophore in visual processes.

The conversion of retinol to retinal occurs through alcohol dehydrogenase enzymes, representing a reversible oxidation reaction. Further oxidation to **retinoic acid** by aldehyde dehydrogenases is irreversible and yields the form responsible for gene regulation and cellular differentiation. The visual cycle involves the isomerization of 11-cis-retinal to all-trans-retinal upon light absorption, triggering the cascade of events leading to visual perception.

Vitamin A deficiency manifests primarily as **night blindness** (nyctalopia) due to impaired rhodopsin regeneration in rod cells. Severe deficiency leads to **xerophthalmia**, characterized by dryness and thickening of conjunctival and corneal tissues, potentially progressing to irreversible blindness. The World Health Organization estimates that vitamin A deficiency affects millions of children globally, making it a significant public health concern in developing nations.

Thiamine (Vitamin B_1): Coenzyme Function and Beriberi

Thiamine consists of a pyrimidine ring connected to a thiazole ring through a methylene bridge. The biologically active form, **thiamine pyrophosphate** (TPP), serves as a coenzyme in carbohydrate metabolism, particularly in the pyruvate dehydrogenase complex and α-ketoglutarate dehydrogenase reactions. The thiazole ring nitrogen can stabilize carbanion intermediates, facilitating carbon-carbon bond cleavage and formation reactions.

The mechanism of TPP action involves the formation of a **carbanion** at the C-2 position of the thiazole ring, which attacks the carbonyl carbon of α-keto acids. This reaction forms a covalent intermediate that undergoes decarboxylation, with the thiazole ring stabilizing the resulting carbanion through resonance.

Thiamine deficiency results in **beriberi**, a condition historically prevalent among populations consuming polished rice as a dietary staple. **Dry beriberi** affects the peripheral nervous system, causing symmetrical sensory and motor deficits, while **wet beriberi** involves cardiovascular complications including high-output heart failure and peripheral edema.

Ascorbic Acid (Vitamin C): Antioxidant Mechanisms

Ascorbic acid possesses a unique enediol structure that readily undergoes oxidation to form **dehydroascorbic acid**. This reversible oxidation-reduction system underlies vitamin C's role as a powerful antioxidant and its function as a cofactor in collagen synthesis. The molecule can donate two electrons sequentially, first forming the semidehydroascorbate radical, then dehydroascorbic acid.

The antioxidant mechanism involves the donation of hydrogen atoms to neutralize free radicals, with ascorbic acid being regenerated through reduction by glutathione or NADH-dependent systems. This cycling mechanism allows vitamin C to function catalytically in antioxidant protection.

Scurvy, the classic vitamin C deficiency disease, results from impaired collagen synthesis due to the inability to hydroxylate proline and lysine residues. Clinical manifestations include bleeding gums, petechial hemorrhages, delayed wound healing, and in severe cases, tooth loss and bone abnormalities.

Vitamin D: Photochemical Synthesis and Calcium Homeostasis

Vitamin D represents a group of secosteroids that function as hormones rather than traditional vitamins. **Cholecalciferol**(vitamin D_3) is synthesized in the skin through the photochemical conversion of 7-dehydrocholesterol upon exposure to ultraviolet B radiation. This process involves the breaking of the B ring of the steroid nucleus, creating the characteristic secosteroid structure.

The activation pathway involves two successive hydroxylation reactions: 25-hydroxylation in the liver by 25-hydroxylase enzymes, followed by 1α-hydroxylation in the kidneys by 1α-hydroxylase to produce **calcitriol** (1,25-dihydroxyvitamin D_3), the most potent vitamin D metabolite.

$$\text{Cholesterol} \rightarrow \text{7-dehydrocholesterol} \rightarrow\text{UV-B}\rightarrow \text{Vitamin } D_3 \rightarrow\text{liver}\rightarrow 25(OH)D_3 \rightarrow\text{kidney}\rightarrow 1,25(OH)_2D_3 \quad (3.4)$$

Vitamin D deficiency causes **rickets** in children and **osteomalacia** in adults, both characterized by defective bone mineralization. The molecular basis involves reduced intestinal calcium absorption and impaired bone matrix calcification.

Vitamin E: Tocopherol Antioxidant System

Vitamin E comprises eight structurally related compounds: four tocopherols (α, β, γ, δ) and four tocotrienols. **α-Tocopherol** exhibits the highest biological activity and serves as the primary vitamin E form in human tissues. The molecule contains a chromanol ring system with a phytyl side chain, providing both antioxidant activity and membrane association properties.

The antioxidant mechanism involves the donation of the phenolic hydrogen to lipid peroxyl radicals, terminating lipid peroxidation chain reactions. The resulting tocopheroxyl radical is relatively stable due to resonance stabilization and can be regenerated through reduction by ascorbic acid or coenzyme Q.

Vitamin Stability During Processing and Storage

The chemical stability of vitamins during food processing and storage represents a critical concern in nutrition and food technology. **Heat sensitivity** varies dramatically among vitamins, with thiamine and vitamin C being particularly labile, while fat-soluble vitamins generally show greater thermal stability.

Oxidative degradation affects many vitamins, particularly those containing unsaturated bonds or easily oxidizable functional groups. Ascorbic acid oxidation follows complex kinetics involving metal catalysis, with copper and iron ions significantly accelerating decomposition rates. The degradation can be represented by:

Ascorbic acid + O_2 →metal catalyst→ Dehydroascorbic acid + H_2O_2 (3.5)

Light exposure causes photodegradation of vitamins, particularly riboflavin and vitamin A. Riboflavin photolysis produces reactive oxygen species that can degrade other vitamins, creating a cascade effect in vitamin destruction.

pH effects significantly influence vitamin stability, with most vitamins showing optimal stability within specific pH ranges. Thiamine stability decreases dramatically above pH 6, while ascorbic acid shows greater stability under acidic conditions.

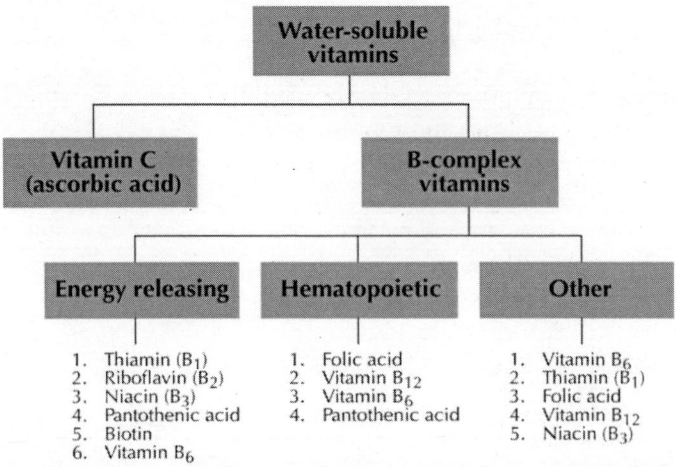

Fig. 13: Chemical structures of representative water-soluble and fat-soluble vitamins showing key functional groups responsible for their biological activities

3.3 Pharmaceutical Chemistry of Drugs

Structure-Activity Relationships in Drug Design

Pharmaceutical chemistry represents the scientific discipline that combines chemistry and pharmacology to design, synthesize, and develop pharmaceutical agents. The foundation of modern drug design rests on **structure-activity relationships (SAR)**, which correlate molecular structure with biological activity. This relationship enables medicinal chemists to optimize drug properties systematically, including potency, selectivity, and pharmacokinetic parameters.

The evolution of pharmaceutical chemistry from empirical observations to rational drug design reflects advances in molecular biology, structural biology, and computational chemistry. **Paul Ehrlich's** concept of the "magic bullet" in the early 1900s laid the conceptual foundation for targeted drug therapy, while the development of sulfonamides by **Gerhard Domagk** demonstrated the potential for synthetic antimicrobial agents.

Modern drug discovery employs **high-throughput screening**, **combinatorial chemistry**, and **computer-aided drug design** to identify and optimize lead compounds. The integration of genomics and proteomics has enabled **personalized medicine** approaches, where drug selection and dosing are tailored to individual genetic profiles.

Aspirin: Anti-inflammatory Mechanisms and Synthesis

Aspirin (acetylsalicylic acid) represents one of the most successful pharmaceutical agents in history, demonstrating multiple therapeutic applications including analgesic, antipyretic, anti-inflammatory, and cardioprotective effects. The molecule consists of a benzoic acid derivative with an acetyl substituent ortho to the carboxyl group, creating a structure capable of irreversible enzyme inhibition.

The **mechanism of action** involves the irreversible acetylation of **Serine-530** in the active site of cyclooxygenase enzymes (COX-1 and COX-2). This covalent modification prevents the conversion of arachidonic acid to prostaglandin H_2, the precursor of various inflammatory mediators including prostaglandins, prostacyclin, and thromboxanes.

The acetylation reaction proceeds through nucleophilic attack by the serine hydroxyl group on the acetyl carbonyl carbon of aspirin:

Aspirin + Ser-530-COX → Acetyl-Ser-530-COX + Salicylic acid (3.6)

Aspirin synthesis can be accomplished through the acetylation of salicylic acid using acetic anhydride in the presence of an acid catalyst, typically phosphoric acid or sulfuric acid. This reaction represents a classic example of **nucleophilic acyl substitution**, where the phenolic hydroxyl group of salicylic acid attacks the carbonyl carbon of acetic anhydride.

The industrial synthesis requires careful control of reaction conditions to minimize side reactions and maximize yield. Temperature control prevents decomposition of the acetyl donor, while the acid catalyst activates the acetic anhydride toward nucleophilic attack.

Paracetamol: Hepatotoxicity and Metabolic Pathways

Paracetamol (acetaminophen) provides analgesic and antipyretic effects through selective inhibition of cyclooxygenase enzymes in the central nervous system. Unlike aspirin, paracetamol shows minimal anti-inflammatory activity in peripheral tissues, reflecting its different mechanism of COX inhibition and tissue distribution properties.

The molecule's structure features an acetamide group para to a phenolic hydroxyl, creating a compound with optimal balance between potency and toxicity within therapeutic dose ranges. However, paracetamol demonstrates a narrow **therapeutic index**, with hepatotoxicity occurring at doses only moderately above therapeutic levels.

Metabolic activation of paracetamol occurs primarily through **cytochrome P450** enzymes, particularly CYP2E1, which converts approximately 5% of the administered dose to the highly reactive metabolite **N-acetyl-p-benzoquinone imine (NAPQI)**. Under normal conditions, NAPQI is rapidly detoxified through conjugation with glutathione.

Paracetamol hepatotoxicity results from glutathione depletion during overdose situations, allowing NAPQI to bind covalently to cellular proteins and initiate oxidative stress cascades. This mechanism explains the effectiveness of **N-acetylcysteine** as an antidote, as it replenishes glutathione stores and provides alternative sulfhydryl groups for NAPQI conjugation.

Penicillin: β-Lactam Chemistry and Resistance Mechanisms

Penicillin represents the first widely used antibiotic and established the foundation for modern antimicrobial chemotherapy. The molecule contains a β-lactam ring fused to a thiazolidine ring, creating a bicyclic system known as the **penam nucleus**. The β-lactam ring exhibits significant ring strain, making it susceptible to nucleophilic attack by bacterial enzymes.

The **mechanism of antibacterial action** involves the irreversible acylation of **transpeptidase enzymes** responsible for bacterial cell wall synthesis. These enzymes, also known as **penicillin-binding proteins (PBPs)**, catalyze the cross-linking of peptidoglycan chains that provide structural integrity to bacterial cell walls.

The acylation reaction proceeds through nucleophilic attack by a serine residue in the enzyme active site on the β-lactam carbonyl carbon, forming a stable acyl-enzyme intermediate that prevents normal substrate binding and catalysis.

Bacterial resistance to penicillin primarily occurs through the production of **β-lactamase enzymes** that hydrolyze the β-lactam ring, destroying the antibiotic's activity. These enzymes have evolved to efficiently catalyze the hydrolysis reaction:

Penicillin + H_2O →β-lactamase→ Penicilloic acid (inactive) (3.7)

The development of β-lactamase inhibitors such as **clavulanic acid** has provided a strategy to overcome this resistance mechanism by irreversibly inactivating the β-lactamase enzymes.

Pharmacokinetics and Drug Metabolism

Pharmacokinetics describes the time course of drug absorption, distribution, metabolism, and excretion (ADME), providing quantitative frameworks for understanding drug behavior in biological systems. These processes determine the concentration of active drug at target sites and influence both therapeutic efficacy and toxicity potential.

Drug absorption depends on molecular properties including lipophilicity, molecular size, charge state, and dissolution characteristics. The **Henderson-Hasselbalch equation** predicts the fraction of drug in ionized versus non-ionized forms at physiological pH:

$$pH = pK_a + \log([A^-]/[HA]) \quad (3.8)$$

First-pass metabolism in the liver can significantly reduce the bioavailability of orally administered drugs, particularly those that are substrates for cytochrome P450 enzymes. This phenomenon explains why some drugs require intravenous administration to achieve therapeutic plasma concentrations.

Phase I metabolism involves oxidation, reduction, and hydrolysis reactions that typically introduce or expose polar functional groups. **Phase II metabolism** involves conjugation reactions with endogenous compounds such as glucuronic acid, sulfate, or glutathione, generally producing water-soluble metabolites that are readily excreted.

Target Pathways and Molecular Mechanisms

Modern pharmaceutical research emphasizes **target-based drug design**, where specific proteins or pathways associated with disease states are identified and validated as therapeutic targets. This approach has led to the development of highly selective drugs with improved efficacy and reduced side effects.

Enzyme inhibition represents a major mechanism of drug action, with drugs designed to bind specifically to enzyme active sites or allosteric sites. **Competitive inhibition** involves drugs that compete with natural substrates for binding, while **non-competitive inhibition** involves binding to sites distinct from the substrate binding site.

Receptor modulation encompasses drugs that act as agonists, antagonists, or partial agonists at specific receptor sites. The development of selective serotonin reuptake inhibitors (SSRIs) for depression treatment exemplifies how understanding neurotransmitter pathways can lead to targeted therapeutic interventions.

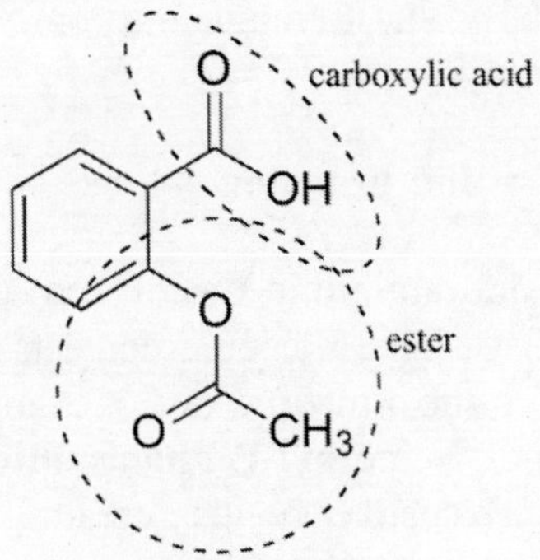

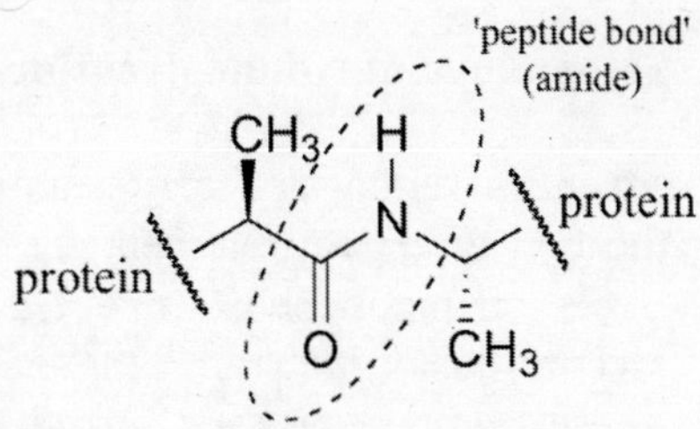

Fig. 14: Three-dimensional structures of aspirin, paracetamol, and penicillin showing key functional groups and their interactions with target enzymes

3.4 Antibiotics and Antiparasitic Agents

Antibiotic Classification and Mechanisms

Antibiotics represent a diverse class of natural and synthetic compounds that exhibit selective toxicity against bacterial pathogens while minimally affecting host cells. The classification of antibiotics is based on their chemical structure, mechanism of action, and spectrum of activity. Understanding these relationships is crucial for appropriate clinical use and combating the growing problem of **antimicrobial resistance**.

The major classes of antibiotics include **β-lactams**, **aminoglycosides**, **macrolides**, **tetracyclines**, **chloramphenicol**, and **quinolones**. Each class targets specific bacterial cellular processes, including cell wall synthesis, protein synthesis, DNA replication, and membrane integrity. The selective toxicity of antibiotics exploits fundamental differences between bacterial and eukaryotic cellular processes.

Bactericidal antibiotics kill bacteria directly, while **bacteriostatic antibiotics** inhibit bacterial growth and reproduction, relying on host immune systems to eliminate the pathogens. This distinction has important clinical implications, particularly in immunocompromised patients who may require bactericidal therapy.

Streptomycin: Aminoglycoside Structure and Function

Streptomycin was the first aminoglycoside antibiotic discovered and the first effective treatment for tuberculosis. The molecule consists of three components: **streptidine** (a substituted cyclohexane), **streptobiosamine** (an aminosugar), and **N-methyl-L-glucosamine**. These components are linked through glycosidic bonds, creating a highly polar, polycationic molecule at physiological pH.

The **mechanism of action** involves binding to the 16S ribosomal RNA component of the bacterial 30S ribosomal subunit. This binding interferes with **protein synthesis** by causing misreading of mRNA codons and preventing the normal translocation of tRNA during translation. The result is the production of aberrant proteins that can damage bacterial cell membranes and other cellular structures.

Streptomycin resistance develops through several mechanisms, including enzymatic modification by aminoglycoside-modifying enzymes, alteration of ribosomal binding sites through mutation, and reduced drug uptake. The rapid development of resistance to streptomycin monotherapy led to the adoption of combination chemotherapy for tuberculosis treatment.

The **clinical applications** of streptomycin have evolved significantly since its discovery. While it remains an important second-line agent for tuberculosis treatment, its use has been limited by **ototoxicity** and **nephrotoxicity**. These adverse effects result from drug accumulation in inner ear and kidney tissues, where the polar nature of aminoglycosides prevents efficient elimination.

Chloramphenicol: Broad Spectrum Activity and Toxicity

Chloramphenicol represents a unique antibiotic with an unusual chemical structure containing a **nitrobenzene ring**, **dichloroacetamide group**, and **propanediol backbone**. This structure enables the molecule to penetrate biological membranes effectively, including the blood-brain barrier, making it valuable for treating central nervous system infections.

The **mechanism of action** involves binding to the **50S ribosomal subunit** at the peptidyl transferase center, preventing peptide bond formation during protein synthesis. This binding is reversible, making chloramphenicol primarily bacteriostatic. The drug shows excellent activity against both gram-positive and gram-negative bacteria, as well as some atypical organisms.

Chloramphenicol toxicity includes two distinct mechanisms: **dose-related bone marrow suppression** and **idiosyncratic aplastic anemia**. The dose-related toxicity results from mitochondrial protein synthesis inhibition in bone marrow cells, as mammalian mitochondrial ribosomes resemble bacterial ribosomes more closely than cytoplasmic ribosomes.

The rare but potentially fatal **aplastic anemia** occurs unpredictably and may be related to the nitrobenzene moiety's metabolism to reactive intermediates that bind to cellular macromolecules. This serious adverse effect has significantly limited chloramphenicol's clinical use in developed countries.

Antimicrobial Resistance: Molecular Mechanisms

Antimicrobial resistance represents one of the most serious threats to global public health, with resistant infections causing increased morbidity, mortality, and healthcare costs. The development of resistance involves complex molecular mechanisms that enable bacteria to survive antibiotic exposure.

Enzymatic inactivation represents a major resistance mechanism, exemplified by β-lactamase enzymes that hydrolyze β-lactam antibiotics. **Extended-spectrum β-lactamases (ESBLs)** have evolved to hydrolyze newer β-lactam antibiotics, while **carbapenemases** can inactivate even the most potent β-lactam agents.

Target modification involves mutations that alter antibiotic binding sites while preserving essential cellular functions. Examples include mutations in ribosomal RNA that reduce aminoglycoside binding and modifications in penicillin-binding proteins that decrease β-lactam affinity.

Efflux pumps actively transport antibiotics out of bacterial cells, reducing intracellular drug concentrations below effective levels. These pumps can be specific for single drugs or exhibit broad substrate specificity, conferring multidrug resistance.

Chloroquine: Antimalarial Mechanisms and Resistance

Chloroquine served as the primary antimalarial agent for several decades and remains important for treating malaria in areas where resistance has not developed. The molecule is a **4-aminoquinoline** derivative with a **diethylamino side chain** that confers its distinctive pharmacological properties.

The **antimalarial mechanism** involves interference with **heme detoxification** in malaria parasites. During their intraerythrocytic stage, *Plasmodium* parasites digest hemoglobin, releasing toxic heme groups that must be sequestered to prevent cellular damage. The parasites convert heme to an inert crystalline form called **hemozoin** (malaria pigment).

Chloroquine disrupts this process by binding to heme and preventing its crystallization, leading to heme accumulation and parasite death:

$$\text{Heme} + \text{Chloroquine} \rightarrow \text{Heme-Chloroquine complex (toxic)} \qquad (3.9)$$

Chloroquine resistance in *P. falciparum* primarily results from mutations in the **P. falciparum chloroquine resistance transporter (PfCRT)** located in the parasite's digestive vacuole. These mutations alter the transport properties of the protein, allowing chloroquine efflux from the vacuole and reducing drug concentration at the site of action.

Mebendazole: Antiparasitic Mechanisms and Applications

Mebendazole belongs to the **benzimidazole** class of antiparasitic agents and demonstrates broad-spectrum activity against intestinal helminths. The molecule's structure contains a **benzimidazole ring** with **methyl carbamate** and **benzoyl** substituents that are essential for antiparasitic activity.

The **mechanism of action** involves binding to β-tubulin subunits in parasite cells, preventing **microtubule polymerization** and disrupting cellular processes including **mitosis**, **transport functions**, and **maintenance of cell shape**. This mechanism selectively affects parasites because their β-tubulin differs significantly from mammalian β-tubulin.

Mebendazole exhibits **selective toxicity** due to its high affinity for parasite β-tubulin compared to mammalian tubulin. Additionally, the drug is poorly absorbed from the gastrointestinal tract, resulting in high local concentrations in the intestine where many target parasites reside, while minimizing systemic exposure.

Clinical applications include treatment of **ascariasis**, **hookworm infection**, **trichuriasis**, and **pinworm infection**. The drug's broad spectrum and favorable safety profile make it suitable for mass drug administration programs in endemic areas.

Drug Resistance in Parasitic Diseases

Parasitic drug resistance poses significant challenges for global health programs, particularly in tropical diseases affecting resource-limited settings. The mechanisms of resistance in parasites often parallel those seen in bacteria but may involve additional complexity due to parasites' more complex life cycles.

Genetic factors contributing to resistance include mutations in target proteins, changes in drug transport systems, and alterations in drug metabolism. The **selection pressure** imposed by widespread drug use creates environments favoring resistant parasite strains.

Cross-resistance between related drugs can limit therapeutic options, as seen with chloroquine and related quinoline antimalarials. Understanding resistance mechanisms enables the development of combination therapies and new drugs designed to overcome existing resistance.

Table 2: Summary of Major Antibiotic Classes and Their Resistance Mechanisms

Antibiotic Class	Representative Drug	Target	Primary Resistance Mechanism
β-lactams	Penicillin	Cell wall synthesis	β-lactamase production

Aminoglycosides	Streptomycin	Protein synthesis (30S)	Enzymatic modification
Phenicols	Chloramphenicol	Protein synthesis (50S)	Acetyltransferase enzymes
Quinolones	Ciprofloxacin	DNA gyrase	Target site mutations

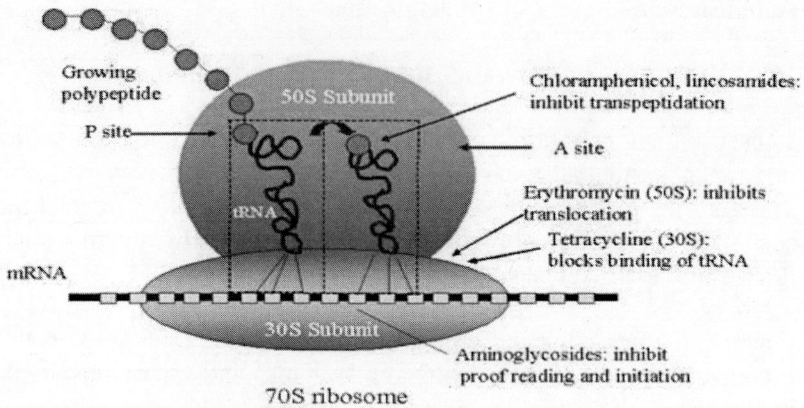

Fig. 15: Comparative structures of streptomycin and chloramphenicol highlighting their distinct mechanisms of ribosomal binding and protein synthesis inhibition

3.5 Oils, Fats, and Soap Chemistry

Lipid Classification and Structural Diversity

Lipids constitute a chemically diverse group of biological molecules unified by their predominantly hydrophobic character and solubility in nonpolar solvents. This class encompasses **triglycerides** (fats and oils), **phospholipids**, **steroids**, and **waxes**, each serving distinct biological functions ranging from energy storage to membrane structure and signaling.

Triglycerides represent the most abundant lipids in both biological systems and commercial applications. These molecules consist of

glycerol esterified with three **fatty acid** chains, creating amphiphilic compounds with varying physical properties determined by fatty acid composition. The systematic name **triacylglycerols** more accurately reflects their chemical structure, emphasizing the acyl linkages formed through esterification reactions.

The **fatty acid composition** of triglycerides determines their physical properties, nutritional value, and industrial applications. **Saturated fatty acids** contain no double bonds and adopt extended conformations that allow close packing, resulting in higher melting points. **Unsaturated fatty acids** contain one or more double bonds, typically in the cis-configuration, which introduce molecular kinks that prevent efficient packing and lower melting points.

Essential fatty acids, including **linoleic acid** (18:2, ω-6) and **α-linolenic acid** (18:3, ω-3), cannot be synthesized by mammalian systems and must be obtained through dietary sources. These polyunsaturated fatty acids serve as precursors for important signaling molecules including **prostaglandins**, **leukotrienes**, and **specialized pro-resolving mediators**.

Rancidity: Chemical Mechanisms and Detection

Rancidity represents the deterioration of fats and oils through chemical reactions that produce objectionable odors, flavors, and potentially harmful compounds. Two primary mechanisms contribute to rancidity: **hydrolytic rancidity** and **oxidative rancidity**, each involving distinct chemical pathways and requiring different prevention strategies.

Hydrolytic rancidity results from the hydrolysis of ester bonds in triglycerides, liberating free fatty acids that contribute to undesirable flavors and odors. This process can occur through enzymatic action by **lipases** or through non-enzymatic acid- or base-catalyzed hydrolysis. The general hydrolysis reaction can be represented as:

$$\text{Triglyceride} + 3H_2O \rightarrow \text{Glycerol} + 3\ \text{Fatty acids} \quad (3.10)$$

Oxidative rancidity involves the **autoxidation** of unsaturated fatty acids through a complex free radical chain reaction. This process begins with the abstraction of hydrogen atoms from methylene groups adjacent to double bonds, forming **lipid radicals** that readily react with molecular oxygen to produce **lipid peroxyl radicals**.

The **autoxidation mechanism** proceeds through three distinct phases: **initiation**, **propagation**, and **termination**. During initiation, lipid radicals (L•) form through hydrogen abstraction by initiating species such as metal ions or existing radicals:

$$LH \rightarrow L\bullet + H\bullet \quad (3.11)$$

The propagation phase involves rapid chain reactions where lipid radicals react with oxygen to form peroxyl radicals (LOO•), which then abstract hydrogen from additional lipid molecules:

$$L\bullet + O_2 \rightarrow LOO\bullet \quad (3.12)$$

$$LOO\bullet + LH \rightarrow LOOH + L\bullet \quad (3.13)$$

Detection methods for rancidity include both chemical and sensory approaches. The **peroxide value** measures primary oxidation products (hydroperoxides) through iodometric titration, providing an early indicator of oxidative changes. The **thiobarbituric acid test** detects secondary oxidation products, particularly **malondialdehyde**, offering insight into advanced oxidation stages.

Acid value determination quantifies free fatty acids present in oils and fats, serving as an indicator of hydrolytic rancidity. The acid value is defined as the milligrams of potassium hydroxide required to neutralize free acids in one gram of sample:

$$\text{Acid Value} = (\text{Volume of KOH} \times \text{Normality} \times 56.1) / \text{Sample weight} \quad (3.14)$$

Antioxidant Systems and Preservation Strategies

Antioxidants prevent or delay oxidative rancidity by interrupting free radical chain reactions through various mechanisms. **Primary antioxidants** donate hydrogen atoms or electrons to free radicals, converting them to stable products. **Secondary antioxidants** work through indirect mechanisms such as metal chelation or oxygen scavenging.

Natural antioxidants include **tocopherols** (vitamin E), **ascorbic acid** (vitamin C), and various **phenolic compounds**found in plant materials. **α-Tocopherol** effectively terminates lipid peroxidation chains by donating its phenolic hydrogen to peroxyl radicals, forming stable products and regenerating through interaction with ascorbic acid.

Synthetic antioxidants such as **butylated hydroxytoluene (BHT)** and **butylated hydroxyanisole (BHA)** provide cost-effective protection for commercial food products. These compounds function through similar radical-scavenging mechanisms but may raise safety concerns at high concentrations.

Synergistic effects occur when combinations of antioxidants provide greater protection than individual compounds. The combination of tocopherols with ascorbic acid exemplifies this synergy, where ascorbic acid regenerates tocopherol radicals, effectively amplifying the antioxidant capacity of the system.

Adulteration Detection in Edible Oils

Oil adulteration represents a significant concern in food safety and economic fraud, involving the intentional addition of lower-cost oils or non-edible substances to higher-value products. Detection methods must be capable of identifying both **quantitative adulteration** (dilution with cheaper oils) and **qualitative adulteration** (addition of non-edible substances).

Physical methods for adulteration detection include **density measurement**, **refractive index determination**, and **viscosity assessment**. These properties reflect the overall fatty acid composition and can reveal significant adulterations, although they may not detect sophisticated adulterations with oils of similar properties.

Chemical methods provide more specific identification of adulterants. **Fatty acid profiling** using **gas chromatography**can detect unusual fatty acid patterns indicative of specific adulterants. **Sterol analysis** exploits the fact that different oil sources contain characteristic sterol compositions that serve as fingerprints for authentication.

Spectroscopic techniques including **infrared spectroscopy** and **nuclear magnetic resonance** provide rapid, non-destructive methods for oil analysis. These techniques can detect adulteration through pattern recognition algorithms that compare sample spectra to authentic reference materials.

DNA-based methods offer highly specific identification of oil sources, particularly valuable for detecting adulteration with genetically modified oils or oils from protected geographical origins. **Polymerase chain reaction (PCR)** techniques can amplify species-specific DNA sequences even in processed oils.

Saponification: Chemistry and Industrial Applications

Saponification represents the alkaline hydrolysis of triglycerides to produce **glycerol** and **carboxylate salts** (soaps). This reaction has been utilized for soap production for millennia and remains fundamental to understanding surfactant chemistry and lipid analysis.

The **saponification reaction** proceeds through nucleophilic attack by hydroxide ions on the carbonyl carbon of ester bonds, forming tetrahedral intermediates that decompose to yield carboxylate anions and glycerol:

$$\text{R-COO-R'} + \text{OH}^- \rightarrow \text{R-COO}^- + \text{R'-OH} \quad (3.15)$$

For triglycerides, the complete saponification requires three moles of base per mole of triglyceride:

$$\text{Triglyceride} + 3\text{NaOH} \rightarrow \text{Glycerol} + 3\ \text{Soap molecules} \quad (3.16)$$

Saponification value represents an important analytical parameter, defined as the milligrams of potassium hydroxide required to saponify one gram of fat or oil. This value reflects the average molecular weight of fatty acids in the sample, with higher values indicating shorter-chain fatty acids.

Industrial soap production typically employs **sodium hydroxide** for hard soaps and **potassium hydroxide** for soft soaps and liquid formulations. The choice of base affects both the physical properties and solubility characteristics of the resulting products.

Hot process and **cold process** methods represent the two primary approaches to soap manufacturing. Hot process methods involve heating reactants to accelerate saponification, while cold process methods rely on the heat generated by the exothermic saponification reaction itself.

Surfactant Chemistry and Detergent Formulations

Surfactants (surface-active agents) represent a broad class of compounds that reduce surface tension at interfaces, enabling applications in cleaning, emulsification, and solubilization. The fundamental structure consists of both **hydrophilic** (water-loving) and **hydrophobic** (water-repelling) regions within the same molecule.

Critical micelle concentration (CMC) represents the concentration above which surfactant molecules spontaneously associate to form **micelles**. Below the CMC, surfactants exist primarily as individual molecules, while above the CMC, additional surfactant molecules form micelles rather than remaining in solution as monomers.

Anionic surfactants include traditional soaps as well as synthetic detergents such as **sodium dodecyl sulfate** and **linear alkylbenzene sulfonates**. These compounds carry negative charges on their hydrophilic heads and demonstrate excellent cleaning properties in alkaline conditions.

Nonionic surfactants such as **polyethylene glycol ethers** and **alkyl polyglucosides** contain uncharged polar head groups. These surfactants show reduced sensitivity to water hardness and pH variations, making them valuable for formulations requiring stability under diverse conditions.

Cationic surfactants bear positive charges and find applications primarily as **fabric softeners** and **disinfectants** due to their ability to bind to negatively charged surfaces such as fabrics and bacterial cell walls.

Biodegradability and Environmental Considerations

Biodegradability of surfactants represents a critical environmental consideration, as these compounds are designed to be released into aquatic systems after use. The molecular structure significantly influences the rate and extent of biodegradation, with **linear alkyl chains** generally showing superior biodegradability compared to **branched structures**.

Primary biodegradation involves the destruction of the surfactant's surface-active properties through modification of the hydrophilic head group. **Ultimate biodegradation** requires complete mineralization to carbon dioxide, water, and inorganic salts.

Aerobic biodegradation typically proceeds through **β-oxidation** of alkyl chains, while **anaerobic conditions** may lead to different degradation pathways and potentially persistent metabolites. **Standardized testing protocols** such as the **OECD Ready Biodegradability Test** provide frameworks for evaluating environmental fate.

Linear alkylbenzene sulfonates (LAS) replaced **branched alkylbenzene sulfonates** in detergent formulations specifically to improve biodegradability. This transition demonstrates how environmental considerations can drive innovations in surfactant chemistry.

Aquatic toxicity considerations include both acute effects on individual organisms and chronic effects on ecosystem function. **Bioaccumulation potential** must also be evaluated, particularly for surfactants with longer alkyl chains that may partition into lipid-rich tissues.

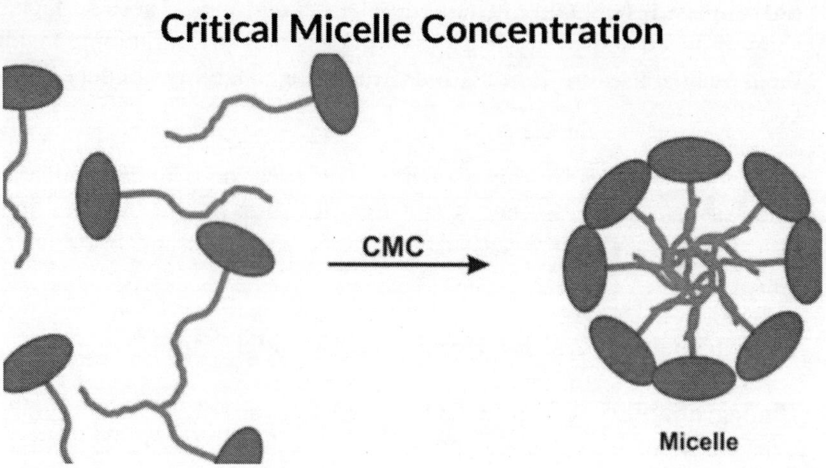

Fig. 16: Schematic representation of micelle formation showing the critical micelle concentration and the arrangement of surfactant molecules in aqueous solution

Key Takeaways

The study of biochemical compounds and medicinal chemistry reveals the intricate relationship between molecular structure and biological function that underlies modern pharmaceutical science and biotechnology. **Carbohydrates** serve as fundamental energy sources and structural components, with their stereochemistry and glycosidic linkages determining both metabolic fate and analytical properties. The phenomenon of **mutarotation** and the development of sophisticated analytical techniques including polarimetry and NMR spectroscopy have enabled detailed characterization of these essential biomolecules.

Vitamin chemistry demonstrates how relatively simple organic molecules can have profound physiological effects, with their water-soluble or fat-soluble nature determining absorption, transport, and storage mechanisms. The **structure-activity relationships** observed in vitamin deficiency diseases provide compelling evidence for the specificity of biochemical processes and highlight the importance of understanding molecular mechanisms in nutrition and public health.

Pharmaceutical chemistry exemplifies the successful application of chemical principles to therapeutic intervention, with drugs like **aspirin**, **paracetamol**, and **penicillin** serving as paradigms for different mechanisms of action. The evolution from empirical drug discovery to rational drug design reflects advances in molecular biology and structural chemistry, enabling the development of increasingly selective and effective therapeutic agents.

Antibiotic and antiparasitic chemistryillustrates both the potential and challenges of antimicrobial chemotherapy. The development of resistance mechanisms underscores the dynamic nature of the relationship between chemicals and biological systems, emphasizing the need for continued innovation in drug discovery and the judicious use of existing antimicrobial agents.

Lipid chemistry encompasses energy storage, membrane structure, and signaling functions, with the chemistry of **oils, fats, and surfactants** bridging biological and industrial applications. Understanding **rancidity mechanisms** and **antioxidant systems** is crucial for food preservation and nutrition, while **surfactant chemistry** enables diverse applications from cleaning products to drug delivery systems.

The integration of these diverse areas of biochemical and medicinal chemistry provides a foundation for understanding how chemical principles govern biological processes and how this knowledge can be applied to improve human health and environmental sustainability. The continued advancement of analytical techniques, computational methods, and synthetic strategies promises to further expand our ability to design and optimize biochemical systems for beneficial applications.

Future developments in this field will likely focus on **personalized medicine** approaches that account for individual genetic variations in drug metabolism, the development of **sustainable chemical processes** for pharmaceutical and consumer product manufacturing, and the design of **biodegradable surfactants** and **green chemistry** approaches that minimize environmental impact while maintaining functional effectiveness.

CHAPTER 4
THERMODYNAMICS, KINETICS, AND GREEN CHEMISTRY

Learning Objectives

By the end of this chapter, students will be able to: • Analyze energy changes in chemical systems using thermodynamic principles and predict reaction spontaneity • Apply phase equilibrium concepts to real-world industrial processes and environmental systems • Calculate reaction rates and determine mechanisms for complex chemical reactions • Evaluate electrochemical energy storage and conversion systems for sustainable applications • Design green polymer synthesis strategies and assess plastic waste management technologies

Introduction

The intersection of thermodynamics, kinetics, and green chemistry represents one of the most critical frontiers in modern chemical science, particularly as global challenges demand sustainable solutions for energy production, environmental protection, and waste management. This chapter explores the fundamental principles that govern chemical transformations and their applications in developing environmentally conscious technologies.

Thermodynamics provides the theoretical framework for understanding energy changes and the driving forces behind chemical reactions, while **kinetics** reveals the pathways and rates at which these transformations occur. The integration of these classical chemical principles with **green chemistry** philosophy has revolutionized how we approach chemical processes, emphasizing atom economy, renewable feedstocks, and environmental compatibility.

The significance of this integration becomes apparent when considering contemporary challenges such as climate change, plastic pollution, and energy security. Traditional industrial chemistry often prioritized efficiency and cost over environmental impact, leading to processes that generated substantial waste and relied on hazardous materials. The modern approach, guided by green chemistry principles established by Paul Anastas and John Warner in the 1990s, seeks to design chemical processes that are inherently safer and more sustainable.

This chapter systematically examines how thermodynamic principles can predict the feasibility of green chemical processes, how kinetic analysis can optimize reaction conditions to minimize waste, and how electrochemical systems can store and convert energy with minimal environmental impact. Furthermore, we explore the thermodynamics of polymer degradation and the kinetics of biodegradation processes, providing essential knowledge for addressing the global plastic waste crisis.

The content progresses from fundamental thermodynamic concepts through advanced applications in sustainable technology, culminating in practical strategies for waste management and circular economy implementation. Each section demonstrates how theoretical principles translate into real-world solutions for environmental and energy challenges.

4.1 Chemical Thermodynamics and Energy Changes

Chemical thermodynamics serves as the foundational science for understanding energy transformations in chemical systems, providing quantitative methods to predict reaction spontaneity, calculate equilibrium positions, and optimize process conditions. The discipline originated from the work of Rudolf Clausius, Josiah Willard Gibbs, and Hermann von Helmholtz in the 19th century, who established the mathematical framework that continues to guide modern chemical engineering and sustainable process design.

Free Energy and Spontaneity

The concept of **free energy** represents one of thermodynamics' most powerful tools for predicting chemical behavior. Gibbs free energy (G) combines enthalpy and entropy effects to provide a single criterion for reaction spontaneity under constant temperature and pressure conditions. This thermodynamic potential becomes particularly valuable in green chemistry applications, where process designers must evaluate multiple reaction pathways to identify the most environmentally favorable routes.

The fundamental relationship for Gibbs free energy change is expressed as:

$$\Delta G = \Delta H - T\Delta S \qquad (4.1)$$

where ΔH represents the enthalpy change, T the absolute temperature, and ΔS the entropy change. When $\Delta G < 0$, the process proceeds spontaneously, while $\Delta G > 0$ indicates non-spontaneous behavior requiring external energy input.

For chemical reactions, the **standard free energy change** ($\Delta G°$) relates to the equilibrium constant through:

$$\Delta G° = -RT \ln K \quad (4.2)$$

This relationship enables prediction of equilibrium positions and calculation of theoretical yields, essential information for designing sustainable processes with minimal waste generation. Large negative $\Delta G°$ values indicate highly favorable reactions that proceed essentially to completion, while small positive values suggest equilibria that may be shifted through creative process design.

The temperature dependence of free energy provides insights into process optimization strategies. The **Gibbs-Helmholtz equation**:

$$(\partial(\Delta G/T)/\partial T)_p = -\Delta H/T^2 \quad (4.3)$$

reveals how equilibrium positions shift with temperature changes, allowing process engineers to identify optimal operating conditions that maximize desired products while minimizing energy consumption.

In sustainable chemistry applications, free energy calculations guide the selection of reaction conditions that minimize environmental impact. For example, catalytic processes often operate at lower temperatures than thermal processes, reducing energy requirements and greenhouse gas emissions. The free energy difference between catalyzed and uncatalyzed pathways quantifies these environmental benefits.

Fugacity and Activity Coefficients in Real Systems

Real chemical systems deviate significantly from ideal behavior, particularly at high pressures, low temperatures, or in concentrated solutions. **Fugacity** and **activity coefficients** provide mathematical tools to account for these deviations, enabling accurate thermodynamic calculations for industrial processes and environmental systems.

Fugacity (f) represents the "effective pressure" of a component in a real gas mixture, defined such that:

$$\mu_i = \mu_i^\circ + RT \ln(f_i/f^\circ_i) \quad (4.4)$$

where μ_i is the chemical potential, μ_i° the standard chemical potential, and f°_i the standard fugacity. The **fugacity coefficient** (φ_i) relates fugacity to pressure through $\varphi_i = f_i/p_i$, providing a measure of deviation from ideal gas behavior.

For liquid solutions, **activity coefficients** (γ_i) similarly correct for non-ideal behavior:

$$a_i = \gamma_i x_i \quad (4.5)$$

where a_i represents activity and x_i the mole fraction. Activity coefficients greater than unity indicate positive deviations from ideality, often associated with unfavorable molecular interactions, while values less than unity suggest negative deviations typically arising from favorable intermolecular forces.

These corrections become crucial in environmental applications where chemical species exist in complex matrices. For instance, the bioavailability of pollutants in soil systems depends on their activity rather than concentration, requiring activity coefficient calculations to predict environmental fate and transport. Similarly, the efficiency of separation processes such as distillation or extraction depends on activity coefficients that determine phase equilibria.

Modern computational methods, including **UNIFAC** (Universal Functional Activity Coefficient) and **COSMO-RS**(Conductor-like Screening Model for Real Solvents), provide predictive tools for estimating activity coefficients in systems where experimental data are unavailable. These methods have proven invaluable for designing green solvents and optimizing separation processes to minimize environmental impact.

Colligative Properties: Osmotic Pressure and Raoult's Law

Colligative properties depend on the number of dissolved particles rather than their chemical identity, providing fundamental insights into solution behavior and practical tools for characterizing complex systems. These properties play crucial roles in biological systems, environmental processes, and industrial applications ranging from water treatment to food preservation.

Raoult's Law describes vapor pressure lowering in ideal solutions:

$$p_i = x_i p^\circ_i \tag{4.6}$$

where p_i represents the partial vapor pressure of component i, x_i its mole fraction in solution, and p°_i the vapor pressure of pure component i. This relationship forms the basis for understanding distillation processes, evaporation rates, and atmospheric chemistry phenomena.

Deviations from Raoult's Law provide information about molecular interactions in solution. Positive deviations ($p_i > x_i p^\circ_i$) indicate weaker intermolecular forces in solution than in pure components, while negative deviations suggest stronger interactions. These insights guide the selection of green solvents that minimize volatility and environmental release.

Osmotic pressure represents perhaps the most practically important colligative property, governing water transport in biological systems and membrane separation processes:

$$\pi = iMRT \tag{4.7}$$

where π is osmotic pressure, i the van't Hoff factor, M the molarity, R the gas constant, and T the absolute temperature. This relationship enables calculation of osmotic pressures in complex systems and design of membrane processes for water treatment and desalination.

Reverse osmosis systems, widely used for water purification, operate by applying pressure exceeding the osmotic pressure of contaminated water, forcing pure water through semi-permeable membranes. The energy requirements for these processes, calculated from thermodynamic principles, determine their economic and environmental feasibility.

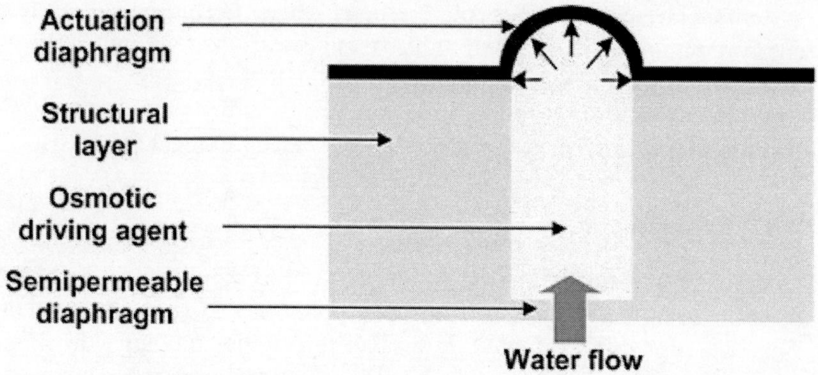

Fig. 17: Schematic representation of osmotic pressure development across a semi-permeable membrane, illustrating the molecular basis for water transport in reverse osmosis systems

Forward osmosis represents an emerging technology that harnesses naturally occurring osmotic pressure gradients for energy-efficient desalination and waste treatment. The thermodynamic driving force eliminates the need for high-pressure pumps, potentially reducing energy consumption by 90% compared to conventional reverse osmosis.

4.2 Phase and Equilibrium Concepts

Phase equilibria govern the distribution of chemical species between different phases, controlling phenomena ranging from atmospheric water cycles to industrial separation processes. Understanding these equilibria provides the foundation for designing efficient, sustainable chemical processes and predicting environmental fate of chemical substances.

Le Chatelier's Principle with Real-World Examples

Le Chatelier's Principle states that systems at equilibrium respond to disturbances by shifting in the direction that opposes the change, providing qualitative predictions for equilibrium behavior under varying conditions. This principle, formulated by Henry Le Chatelier in 1884, continues to guide process optimization and environmental management strategies.

The quantitative foundation for Le Chatelier's Principle emerges from thermodynamic relationships. For a general reaction $aA + bB \rightleftharpoons cC +$

dD, the equilibrium constant varies with temperature according to the **van't Hoff equation**:

$$d \ln K/dT = \Delta H°/RT^2 \tag{4.8}$$

This relationship reveals that exothermic reactions ($\Delta H° < 0$) show decreasing equilibrium constants with increasing temperature, while endothermic reactions exhibit the opposite behavior. Industrial applications leverage this temperature dependence to optimize reaction conditions and product yields.

The **Haber-Bosch process** for ammonia synthesis exemplifies practical application of Le Chatelier's Principle. The reaction $N_2 + 3H_2 \rightleftharpoons 2NH_3$ is exothermic with $\Delta H° = -92$ kJ/mol, suggesting lower temperatures favor ammonia formation. However, reaction rates become prohibitively slow at low temperatures, requiring compromise conditions around 400-500°C with high pressure (150-300 atm) to achieve acceptable rates while maintaining reasonable equilibrium conversion.

Pressure effects follow from the principle that equilibria shift to reduce the number of gas molecules when pressure increases. The Haber process benefits from high pressure because four molecules of reactants produce two molecules of product, representing a 50% volume reduction. This pressure dependence quantifies as:

$$d \ln K/dp = -\Delta V°/RT \tag{4.9}$$

Environmental applications of Le Chatelier's Principle include **carbon sequestration** strategies. Ocean acidification results from CO_2 dissolution forming carbonic acid: $CO_2 + H_2O \rightleftharpoons H_2CO_3 \rightleftharpoons H^+ + HCO_3^-$. Increasing atmospheric CO_2 concentrations drive this equilibrium forward, lowering ocean pH with consequences for marine ecosystems. Understanding these equilibrium shifts enables prediction of environmental impacts and development of mitigation strategies.

Catalytic converter operation in automobiles demonstrates Le Chatelier's Principle in pollution control. The water-gas shift reaction $CO + H_2O \rightleftharpoons CO_2 + H_2$ converts toxic carbon monoxide to less harmful products. High-temperature operation shifts equilibrium toward reactants, but the presence of catalysts enables acceptable conversion rates at moderate temperatures where equilibrium favors products.

Phase Diagrams and Eutectic Systems

Phase diagrams provide graphical representations of phase equilibria, mapping the conditions under which different phases exist and coexist. These diagrams serve as roadmaps for understanding material behavior under varying temperature and pressure conditions, guiding everything from metallurgical processes to pharmaceutical formulation.

The fundamental thermodynamic principle underlying phase diagrams is the **Gibbs Phase Rule**:

$$F = C - P + 2 \tag{4.10}$$

where F represents degrees of freedom, C the number of components, and P the number of phases. This relationship determines how many intensive variables can be independently varied while maintaining equilibrium.

Eutectic systems exhibit particularly important behavior where two or more components form mixtures with minimum melting points. The **eutectic point** represents the composition and temperature at which liquid and multiple solid phases coexist in equilibrium. At this point, the degrees of freedom equal zero ($F = 2 - 3 + 2 = 1$), meaning only temperature or pressure can be independently varied.

The **lead-tin system** historically provided the foundation for understanding eutectic behavior. At the eutectic composition (61.9% tin), the alloy melts at 183°C, significantly below the melting points of pure lead (327°C) or tin (232°C). This principle enables the formulation of solders with precisely controlled melting points for electronic applications.

Modern applications of eutectic systems include **deep eutectic solvents** (DES), which represent sustainable alternatives to conventional organic solvents. These systems typically combine hydrogen bond donors and acceptors in specific ratios that dramatically depress melting points. For example, choline chloride combined with urea in a 1:2 molar ratio forms a liquid at room temperature despite both components being high-melting solids.

Table 3: Common Deep Eutectic Solvent Systems and Their Applications

Hydrogen Bond Acceptor	Hydrogen Bond Donor	Molar Ratio	Melting Point (°C)	Applications
Choline chloride	Urea	1:2	12	Metal extraction, electroplating
Choline chloride	Glycerol	1:2	-40	Biocatalysis, pharmaceuticals
Choline chloride	Malonic acid	1:1	10	Organic synthesis, separations

The thermodynamics of eutectic formation involves **negative deviations** from ideal mixing behavior, often attributed to hydrogen bonding or other specific interactions between components. The **excess Gibbs free energy** of mixing:

$$G^E = RT \Sigma x_i \ln \gamma_i \quad (4.11)$$

becomes strongly negative in eutectic systems, indicating favorable intermolecular interactions that stabilize the liquid phase relative to pure components.

Environmental applications of phase diagram principles include understanding the fate of **salt contamination** in groundwater systems. The $NaCl-H_2O$ phase diagram predicts freezing point depression in contaminated water, affecting seasonal freeze-thaw cycles and groundwater flow patterns. Similarly, the formation of gas hydrates in deep-sea environments depends on pressure-temperature conditions that can be predicted from phase diagrams.

Partial Molar Properties and Mixing Thermodynamics

Partial molar properties quantify how extensive properties change with composition in multicomponent systems, providing essential tools for understanding mixing processes, separation efficiency, and environmental fate of chemical mixtures. These properties become particularly important in green chemistry applications where minimizing waste requires precise understanding of component behavior in complex mixtures.

The **partial molar volume** ($\bar{V}_i$) represents the most fundamental partial molar property:

$$\bar{V}_i = (\partial V/\partial n_i)T,p,n_j \quad (4.12)$$

This quantity often differs significantly from molar volumes of pure components, reflecting changes in molecular packing and intermolecular interactions upon mixing. Understanding these volume changes enables prediction of density variations in mixed systems and design of separation processes.

Apparent molar properties provide experimentally accessible alternatives to partial molar properties, particularly useful in dilute solutions. The apparent molar volume:

$$\varphi_v = (V - n_1V_1°)/n_2 \quad (4.13)$$

approaches the partial molar volume at infinite dilution, providing insights into ion-solvent interactions and hydration phenomena.

The **excess thermodynamic properties** quantify deviations from ideal mixing behavior:

$$A^E = A^{real} - A^{ideal} \quad (4.14)$$

where A represents any extensive property. Excess properties provide molecular-level insights into intermolecular interactions and enable prediction of phase behavior in complex systems.

Activity-composition relationships derived from excess Gibbs free energy models enable prediction of phase equilibria in multicomponent systems. The **Wilson equation**:

$$\ln \gamma_i = -\ln(\Sigma_j x_j\Lambda_{ij}) + 1 - \Sigma_k(x_k\Lambda_{ki}/\Sigma_j x_j\Lambda_{kj}) \quad (4.15)$$

provides accurate correlations for many liquid mixtures, where Λ_{ij} represents adjustable parameters related to molecular interactions.

Environmental applications of mixing thermodynamics include modeling**contaminant partitioning** between different phases. The distribution of organic pollutants between water and organic phases depends on activity coefficients that reflect molecular interactions. Understanding these interactions enables prediction of bioaccumulation potential and design of remediation strategies.

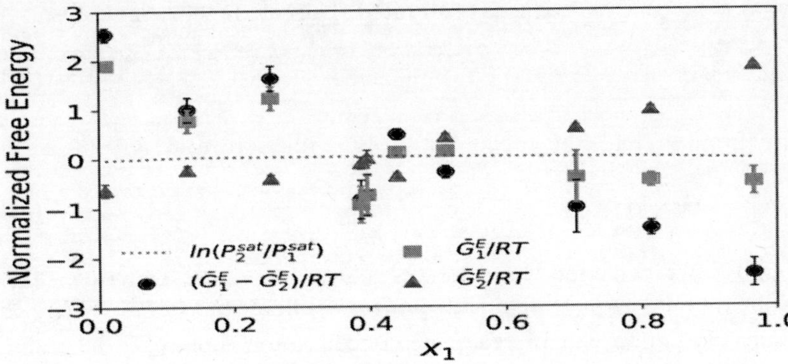

Fig. 18: Excess Gibbs free energy curves for different types of liquid mixtures, illustrating the relationship between molecular interactions and thermodynamic non-ideality

Sustainable solvent design leverages mixing thermodynamics to identify green alternatives to hazardous solvents. **COSMO-RS calculations** predict activity coefficients and solubility parameters for novel solvents, enabling virtual screening of environmentally benign alternatives without extensive experimental work.

4.3 Reaction Kinetics and Mechanisms

Chemical kinetics provides the temporal dimension to chemical understanding, revealing how fast reactions proceed and through what pathways they occur. While thermodynamics determines the feasibility and ultimate equilibrium position of reactions, kinetics governs the rates at which these equilibria are achieved and the mechanisms by which chemical bonds are broken and formed.

Rate Laws for Zero, First, and Second-Order Reactions

Reaction rates fundamentally depend on reactant concentrations, temperature, and the presence of catalysts. The mathematical relationships describing these dependencies, known as **rate laws**, provide quantitative tools for predicting reaction behavior and optimizing process conditions for maximum efficiency and minimal environmental impact.

Zero-order kinetics occur when reaction rates remain independent of reactant concentration, typically observed in heterogeneous catalysis where the catalyst surface is saturated. The rate law takes the form:

$$\text{rate} = k \quad (4.16)$$

Integration yields the concentration-time relationship:

$$[A] = [A]_0 - kt \quad (4.17)$$

Zero-order behavior characterizes many **enzymatic reactions** at high substrate concentrations and **photochemical processes** where light intensity rather than reactant concentration controls the rate. In environmental applications, zero-order kinetics often describe the biodegradation of pollutants in systems where microbial populations are saturated with substrate.

First-order kinetics represent the most commonly encountered rate law, where reaction rate is directly proportional to reactant concentration:

$$\text{rate} = k[A] \quad (4.18)$$

The integrated form provides the exponential decay characteristic of first-order processes:

$$[A] = [A]_0 e^{(-kt)} \quad (4.19)$$

The **half-life** ($t_{1/2}$) for first-order reactions equals ln(2)/k, providing a concentration-independent measure of reaction progress. This characteristic makes first-order kinetics particularly valuable for describing **radioactive decay**, **drug metabolism**, and many **environmental degradation processes**.

Second-order kinetics exhibit more complex behavior, with rates depending on the square of a single reactant concentration or the product of two reactant concentrations. For the case where rate = $k[A]^2$, integration yields:

$$1/[A] = 1/[A]_0 + kt \tag{4.20}$$

The half-life for second-order reactions depends on initial concentration: $t_{1/2} = 1/(k[A]_0)$, making these reactions increasingly slow as they proceed. Many **bimolecular reactions** in solution follow second-order kinetics, including **SN2 substitution reactions** and **radical recombination processes.**

Pseudo-first-order conditions often simplify complex kinetic analysis by maintaining one reactant in large excess. Under these conditions, a bimolecular reaction A + B → products with rate = k[A][B] behaves as if first-order in A when [B] >> [A], with an apparent rate constant k' = k[B]. This approach frequently applies in **biochemical systems** and **environmental remediation** processes.

Table 4: Comparison of Reaction Order Characteristics

Order	**Rate Law**	**Integrated Form**	**Half-Life**	**Linear Plot**
0	k	$[A] = [A]_0 - kt$	$[A]_0/2k$	[A] vs t
1	k[A]	$[A] = [A]_0e^{(-kt)}$	ln(2)/k	ln[A] vs t
2	$k[A]^2$	$1/[A] = 1/[A]_0 + kt$	$1/(k[A]_0)$	1/[A] vs t

Transition State Theory and Activation Energy

Transition State Theory (TST), developed by Henry Eyring, Meredith Evans, and Michael Polanyi in the 1930s, provides a molecular-level understanding of reaction rates by examining the energy pathway connecting reactants to products. This theory forms the foundation for rational catalyst design and process optimization in green chemistry applications.

The central premise of TST assumes that reactions proceed through a **transition state** or **activated complex** that represents the highest energy point along the reaction coordinate. The **activation energy** (Ea) quantifies the energy barrier that must be overcome for reaction to occur, determining the temperature dependence of reaction rates through the **Arrhenius equation**:

$$k = Ae^{(-Ea/RT)} \tag{4.21}$$

where A represents the **pre-exponential factor** related to the frequency of molecular collisions and the probability that collisions result in reaction.

TST provides a more detailed expression for the rate constant:

$$k = (k_BT/h)(K^‡)e^{(-\Delta G^‡/RT)} \tag{4.22}$$

where kB is Boltzmann's constant, h Planck's constant, K‡ the equilibrium constant for transition state formation, and ΔG‡ the **Gibbs activation energy**. This relationship reveals how both enthalpic and entropic factors contribute to reaction rates.

The **activation enthalpy** (ΔH‡) and **activation entropy** (ΔS‡) provide insights into reaction mechanisms:

$$\Delta G^‡ = \Delta H^‡ - T\Delta S^‡ \tag{4.23}$$

Positive activation entropies suggest transition states with increased disorder compared to reactants, often indicating bond-breaking processes, while negative values suggest more ordered transition states typical of association reactions.

Catalysis fundamentally involves providing alternative reaction pathways with lower activation energies. **Homogeneous catalysts** operate in the same phase as reactants, forming temporary bonds that stabilize transition states. **Heterogeneous catalysts** provide surfaces that adsorb reactants, bringing them into close proximity and orienting them favorably for reaction.

Enzymatic catalysis represents nature's solution to achieving high reaction rates under mild conditions. Enzymes typically reduce activation energies by 50-85 kJ/mol, corresponding to rate enhancements of 10^6 to 10^{17}. Understanding enzymatic mechanisms guides the development of **biomimetic catalysts** for sustainable chemical processes.

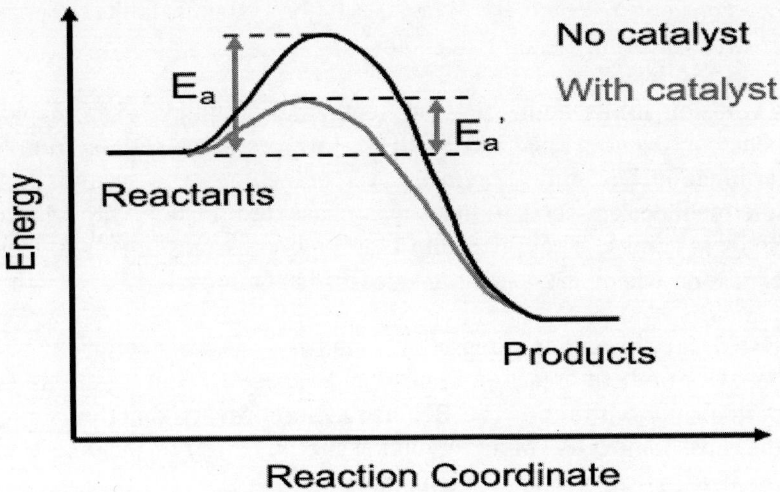

Fig. 19: Energy profile diagram comparing uncatalyzed and catalyzed reaction pathways, illustrating how catalysts reduce activation energy without affecting overall thermodynamics

Branched Chain and Oscillatory Reactions

Branched chain reactions involve propagation steps that produce more radical species than they consume, leading to exponential increases in reaction rates and, under certain conditions, **explosive behavior**. These reactions demonstrate how molecular-level mechanisms determine macroscopic kinetic behavior and highlight the importance of reaction control in industrial safety.

The **hydrogen-oxygen reaction** exemplifies branched chain behavior:

Chain initiation: $H_2 + O_2 \rightarrow 2OH$ Chain propagation: $OH + H_2 \rightarrow H_2O + H$ Branching step: $H + O_2 \rightarrow OH + O$ Second branching: $O + H_2 \rightarrow OH + H$

The branching steps produce two radicals from one, leading to exponential radical concentration growth when the branching rate exceeds termination rates. The **chain branching factor** (φ) quantifies this behavior:

$$\varphi = (kb - kt)/kp \tag{4.24}$$

where kb, kt, and kp represent branching, termination, and propagation rate constants respectively. When $\varphi > 0$, the system exhibits exponential growth potentially leading to explosion.

Explosion limits define pressure-temperature regions where explosive behavior occurs. The **first limit** at low pressures results from wall termination becoming dominant over branching. The **second limit** at intermediate pressures reflects gas-phase termination through three-body reactions. The **third limit** at high pressures involves thermal explosion when heat generation exceeds heat removal.

Oscillatory reactions represent another class of complex kinetic behavior where reaction rates and species concentrations vary periodically with time. The **Belousov-Zhabotinsky reaction** serves as the most famous example, involving the oxidation of malonic acid by bromate in the presence of cerium ions:

$$BrO_3^- + CH_2(COOH)_2 + Ce^{3+} + H^+ \rightarrow \text{oscillating system}$$

The oscillatory behavior arises from **autocatalytic feedback loops** where reaction products influence their own formation rates. Mathematical analysis reveals that oscillations require at least three species and **nonlinear kinetics** with positive and negative feedback mechanisms.

Environmental implications of oscillatory reactions include understanding **atmospheric chemistry oscillations** such as those observed in stratospheric ozone concentrations and **ecological oscillations** in predator-prey systems. These phenomena demonstrate how chemical kinetics influences environmental stability and sustainability.

The **oregonator model** provides a simplified mathematical description of oscillatory behavior:

$$A + Y \rightarrow X + P \tag{4.25a}$$

$$X + Y \rightarrow 2P \tag{4.25b}$$

$$A + X \rightarrow 2X + 2Z \tag{4.25c}$$

$$2X \rightarrow A + P \tag{4.25d}$$

$$B + Z \rightarrow Y \tag{4.25e}$$

This model demonstrates how **autocatalysis** (step c) combined with **inhibition** (step b) generates sustained oscillations under appropriate conditions.

4.4 Electrochemical and Solar Energy Systems

Electrochemical systems represent crucial technologies for sustainable energy storage and conversion, providing pathways to integrate renewable energy sources with existing infrastructure while minimizing environmental impact. These systems harness the principles of electron transfer reactions to achieve energy transformations with high efficiency and controllability.

Galvanic vs Electrolytic Cells

Galvanic cells convert chemical energy directly into electrical energy through spontaneous oxidation-reduction reactions, while **electrolytic cells** use electrical energy to drive non-spontaneous chemical transformations. Understanding the fundamental differences between these systems provides the foundation for designing energy storage and conversion technologies.

In galvanic cells, the **Gibbs free energy change** for the overall cell reaction is negative, indicating spontaneous operation. The maximum electrical work obtainable equals the free energy change:

$$Wmax = -nFE^{\circ} \tag{4.26}$$

where n represents the number of electrons transferred, F the Faraday constant (96,485 C/mol), and E° the **standard cell potential**. The cell potential relates to the equilibrium constant through:

$$E^{\circ} = (RT/nF) \ln K \tag{4.27}$$

The **Nernst equation** describes how cell potential varies with temperature and concentration:

$$E = E^{\circ} - (RT/nF) \ln Q \tag{4.28}$$

where Q represents the reaction quotient. This relationship enables calculation of cell potentials under non-standard conditions and prediction of performance variations during operation.

Electrolytic cells require applied potentials exceeding the thermodynamic minimum to drive reactions forward. The minimum voltage equals the absolute value of the cell potential calculated from standard reduction potentials. Additional **overpotential** compensates for kinetic limitations and resistance losses:

$$E_{applied} = |E°| + \eta_a + \eta_c + iR \qquad (4.29)$$

where η_a and η_c represent anodic and cathodic overpotentials, i the current, and R the cell resistance.

Battery systems combine galvanic and electrolytic operation, discharging as galvanic cells and recharging as electrolytic cells. The **energy efficiency** of charge-discharge cycles depends on the difference between discharge and charge voltages:

$$\eta_{energy} = E_{discharge}/E_{charge} \qquad (4.30)$$

Typical lithium-ion batteries achieve energy efficiencies of 90-95%, making them attractive for grid-scale energy storage applications.

Table 5: Standard Reduction Potentials for Common Battery Reactions

Half-Reaction	E° (V vs SHE)	Application
$Li^+ + e^- \rightarrow Li$	-3.04	Lithium-ion batteries
$Na^+ + e^- \rightarrow Na$	-2.71	Sodium-ion batteries
$Zn^{2+} + 2e^- \rightarrow Zn$	-0.76	Zinc-air batteries
$2H^+ + 2e^- \rightarrow H_2$	0.00	Hydrogen fuel cells
$O_2 + 4H^+ + 4e^- \rightarrow 2H_2O$	+1.23	Oxygen reduction

Fuel Cells: PEMFC and SOFC

Fuel cells represent the most direct method for converting chemical energy to electrical energy, achieving higher efficiencies than combustion-based power generation while producing minimal pollutants. The **theoretical efficiency** of fuel cells approaches 100% because they avoid the Carnot cycle limitations that restrict heat engines.

Proton Exchange Membrane Fuel Cells (PEMFC) operate at relatively low temperatures (60-80°C) using hydrogen fuel and oxygen from air. The overall reaction:

$$H_2 + \frac{1}{2}O_2 \rightarrow H_2O \qquad (4.31)$$

occurs through separate half-reactions at the anode and cathode:

Anode: $H_2 \rightarrow 2H^+ + 2e^-$ Cathode: $\frac{1}{2}O_2 + 2H^+ + 2e^- \rightarrow H_2O$

The **proton exchange membrane** allows proton transport while blocking electron flow, forcing electrons through the external circuit to generate useful work. **Platinum catalysts** facilitate both half-reactions, though their high cost and scarcity motivate research into alternative catalysts.

The **theoretical voltage** of a hydrogen fuel cell equals 1.23 V at standard conditions, calculated from the standard free energy change. However, practical cells deliver 0.6-0.8 V under load due to **activation losses**, **ohmic losses**, and **concentration losses**:

$$E_{cell} = E° - \eta_{activation} - \eta_{ohmic} - \eta_{concentration} \qquad (4.32)$$

Activation losses result from slow electrode kinetics, particularly for oxygen reduction. The **Tafel equation** describes the relationship between overpotential and current density:

$$\eta = a + b \log(i) \qquad (4.33)$$

where a and b are constants dependent on the electrode material and reaction mechanism.

Solid Oxide Fuel Cells (SOFC) operate at high temperatures (800-1000°C) using ceramic electrolytes that conduct oxygen ions rather than protons. This design enables direct utilization of hydrocarbon fuels through **internal reforming**:

$$CH_4 + H_2O \rightarrow CO + 3H_2 \qquad (4.34)$$

$$CO + H_2O \rightarrow CO_2 + H_2 \qquad (4.35)$$

The high operating temperature eliminates the need for precious metal catalysts and enables **cogeneration** of heat and power, achieving overall efficiencies exceeding 80%.

Fuel cell systems require **balance of plant** components including fuel processing, air supply, thermal management, and power conditioning. The system efficiency depends on all components, not just the fuel cell stack. Advanced systems incorporate **heat recovery** and **fuel recycling** to maximize overall efficiency.

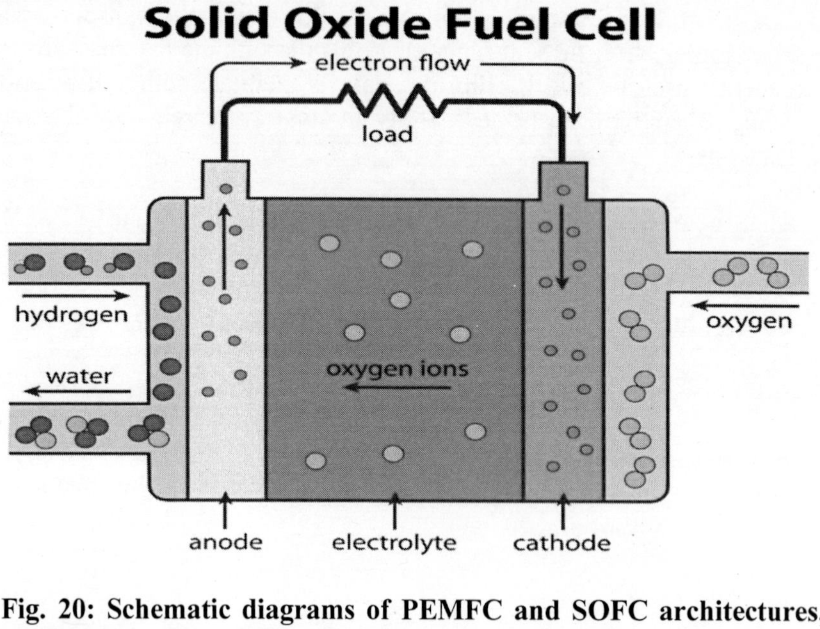

Fig. 20: Schematic diagrams of PEMFC and SOFC architectures, highlighting the different electrolyte materials and operating principles

Energy Storage in Secondary Batteries (Li-ion)

Lithium-ion batteries have revolutionized portable electronics and are transforming transportation and grid storage applications through their combination of high energy density, long cycle life, and operational flexibility. Understanding the electrochemical principles governing lithium-ion operation enables optimization of performance and identification of limitations requiring technological advancement.

The **intercalation mechanism** in lithium-ion batteries involves reversible insertion of lithium ions into host materials without structural breakdown. During discharge, lithium ions move from the **anode** (typically graphite) through the **electrolyte** to the **cathode** (often a lithium metal oxide), while electrons flow through the external circuit to provide useful work.

The **graphite anode** stores lithium through intercalation between graphene layers, forming compounds with stoichiometry up to LiC_6. The reaction during discharge:

$$LiC_6 \rightarrow Li^+ + e^- + C_6 \qquad (4.36)$$

provides a stable, low-potential source of lithium ions. Graphite's layered structure accommodates lithium insertion with minimal volume change, contributing to excellent cycle life.

Cathode materials determine the energy density and operating voltage of lithium-ion cells. **Lithium cobalt oxide**($LiCoO_2$) was the first commercial cathode, providing high energy density but limited thermal stability. The discharge reaction:

$$LiCoO_2 \rightarrow Li_{1-x}CoO_2 + xLi^+ + xe^- \qquad (4.37)$$

typically allows extraction of 50-60% of theoretical lithium content to maintain structural stability.

Lithium iron phosphate ($LiFePO_4$) offers improved safety and cycle life through its **olivine structure** that resists thermal runaway. The **two-phase reaction mechanism**:

$$LiFePO_4 \rightleftharpoons FePO_4 + Li^+ + e^- \qquad (4.38)$$

maintains constant voltage during most of the discharge, simplifying battery management systems.

Electrolyte selection critically influences battery performance, safety, and lifetime. Conventional organic electrolytes combine **lithium salts** (typically $LiPF_6$) with **organic carbonates** (ethylene carbonate, dimethyl carbonate). The electrolyte must provide high ionic conductivity while maintaining chemical stability over the operating voltage range.

Solid-state electrolytes represent emerging technology for enhanced safety and energy density. **Ceramic electrolytes**such as LLZO ($Li_7La_3Zr_2O_{12}$) eliminate flammability risks while enabling lithium metal anodes with theoretical capacity of 3,860 mAh/g compared to graphite's 372 mAh/g.

The **energy density** of lithium-ion cells depends on both voltage and capacity:

$$\text{Energy density} = \text{Voltage} \times \text{Capacity} \tag{4.39}$$

Current commercial cells achieve 150-250 Wh/kg, while research targets exceed 500 Wh/kg through advanced materials and architectures.

Battery degradation occurs through multiple mechanisms including **solid electrolyte interphase** (SEI) growth, **lithium plating**, **cathode material dissolution**, and **electrolyte decomposition**. Understanding these mechanisms enables development of mitigation strategies and accurate lifetime prediction models.

4.5 Green Polymers and Plastic Waste Management

The global plastic crisis demands fundamental changes in how polymers are designed, produced, and managed at end-of-life. **Green polymers** represent materials designed with environmental considerations from synthesis through disposal, incorporating principles of renewable feedstocks, biodegradability, and circular economy thinking. This section examines the scientific principles underlying sustainable polymer technologies and waste management strategies.

Biodegradable Polymer Synthesis

Biodegradable polymers undergo degradation through biological processes, typically involving enzymatic cleavage of polymer chains into smaller molecules that can be metabolized by microorganisms. The design of biodegradable polymers requires understanding the relationship between molecular structure and susceptibility to biological attack.

Hydrolyzable linkages form the foundation of most biodegradable polymers. **Ester bonds**, **amide bonds**, and **anhydride bonds** are susceptible to enzymatic cleavage, while **carbon-carbon bonds** in conventional polyolefins resist biological degradation. The hydrolysis reaction for ester-containing polymers:

$$\text{R-COO-R'} + H_2O \rightarrow \text{R-COOH} + \text{R'-OH} \qquad (4.40)$$

can occur through both chemical and enzymatic pathways, with enzymes providing specificity and enhanced reaction rates under mild conditions.

Polylactic acid (PLA) represents the most commercially successful biodegradable polymer, synthesized from renewable **lactic acid** derived from corn starch or sugarcane. The **ring-opening polymerization** of lactide:

$$\text{n(lactide)} \rightarrow \text{-(CHCHCOo)}_n\text{-} \qquad (4.41)$$

CH_3

produces polymers with controlled molecular weight and stereochemistry. The presence of **chiral centers** influences crystallinity, mechanical properties, and degradation rates.

Polyhydroxyalkanoates (PHAs) are produced directly by microorganisms as intracellular carbon and energy storage materials. **Polyhydroxybutyrate** (PHB) and its copolymers offer complete biodegradability in marine environments, addressing concerns about ocean plastic pollution. The biosynthesis pathway involves:

Acetyl-CoA → Acetoacetyl-CoA → β-Hydroxybutyryl-CoA → PHB

Starch-based polymers combine renewable feedstocks with controlled biodegradability. **Thermoplastic starch** (TPS) is produced by processing starch with plasticizers under heat and shear, disrupting crystalline regions to create thermoplastic materials. **Starch-polymer**

blends improve mechanical properties while maintaining biodegradability.

The **biodegradation kinetics** of polymers follow complex pathways involving multiple steps: • **Surface erosion** where degradation occurs primarily at the polymer surface • **Bulk erosion** where degradation occurs throughout the polymer matrix • **Enzymatic cleavage** of specific bonds by microbial enzymes • **Microbial consumption** of degradation products

Table 6: Biodegradation Times for Common Polymers in Different Environments

Polymer	Composting (weeks)	Marine (months)	Soil (years)
PLA	12-16	6-24	1-2
PHB	4-8	1-6	0.5-1
Starch blend	2-4	1-3	0.25-0.5
Conventional PE	No degradation	No degradation	100-1000

Life Cycle Analysis and Degradation Pathways

Life Cycle Assessment (LCA) provides systematic methodology for evaluating environmental impacts of polymer systems from raw material extraction through end-of-life management. LCA enables comparison of different materials and identification of opportunities for environmental improvement.

The **ISO 14040 framework** defines four phases of LCA: • **Goal and scope definition** establishing the purpose and boundaries of the study • **Inventory analysis** quantifying inputs and outputs throughout the life cycle • **Impact assessment** characterizing potential environmental effects • **Interpretation** drawing conclusions and recommendations

Cradle-to-grave analysis of conventional plastics reveals significant environmental burdens including **fossil fuel depletion**, **greenhouse gas emissions**, and **persistent waste accumulation**. Petroleum-based polymers typically require 1.5-2.0 kg of crude oil per kg of polymer produced, contributing to resource depletion and climate change.

Biopolymer LCA often shows advantages in **renewable resource utilization** and **end-of-life biodegradation** but may have higher impacts in **land use**, **eutrophication**, and **energy consumption** during production. **PLA production** requires significant energy for fermentation and polymerization, though this can be offset by renewable energy sources.

Degradation pathway analysis examines the mechanisms and products of polymer breakdown in different environments. **Aerobic biodegradation** in composting environments proceeds through:

Polymer → Oligomers → Monomers → CO_2 + H_2O + Biomass

Anaerobic degradation in landfills produces different products:

Polymer → Oligomers → Monomers → CH_4 + CO_2 + Biomass

The **CO_2 evolution test** (ASTM D5338) quantifies biodegradation rates by measuring carbon dioxide production during controlled composting. Complete biodegradation requires at least 90% conversion of polymer carbon to CO_2 within 180 days.

Marine degradation presents unique challenges due to lower temperatures, variable oxygen levels, and different microbial communities. **Hydrolysis rates** decrease significantly in seawater compared to terrestrial environments, extending degradation times for even biodegradable polymers.

Photodegradation occurs through **ultraviolet radiation** causing polymer chain scission and oxidation. **Pro-oxidant additives** can accelerate photodegradation, though complete mineralization may require subsequent biological processes.

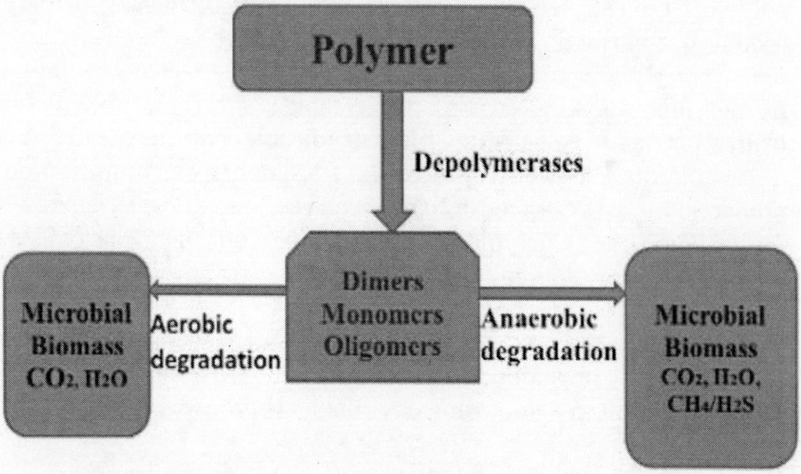

Fig. 21: Schematic representation of polymer degradation pathways in different environmental compartments, showing the influence of oxygen, temperature, and microbial activity

Chemical Strategies for Plastic Recycling

Chemical recycling offers pathways to recover value from plastic waste through molecular-level transformations that regenerate monomers or produce useful chemicals. These processes address limitations of mechanical recycling, which degrades polymer properties through repeated processing cycles.

Depolymerization reverses polymerization reactions to recover original monomers. **PET depolymerization** through **glycolysis** breaks ester bonds using ethylene glycol:

$$-(OCH_2CH_2OOC\text{-}C_6H_4\text{-}CO)_n\text{-} + nHOCH_2CH_2OH \rightarrow \quad (4.42)$$

$$nHOCH_2CH_2OOC\text{-}C_6H_4\text{-}COOH + nHOCH_2CH_2OH$$

The recovered **bis(2-hydroxyethyl) terephthalate** (BHET) can be repolymerized to virgin-quality PET, achieving true circular recycling.

Methanolysis of PET produces **dimethyl terephthalate** (DMT) and ethylene glycol:

$$PET + CH_3OH \rightarrow DMT + \text{ethylene glycol} \quad (4.43)$$

This process operates at higher temperatures (180-280°C) but produces higher-purity products suitable for food-grade applications.

Pyrolysis thermally decomposes polymers in the absence of oxygen, producing mixtures of hydrocarbons that can serve as feedstocks for new polymer production or fuel applications. **Polyolefin pyrolysis** produces complex mixtures of alkanes, alkenes, and aromatics:

$$-(CH_2\text{-}CHR)_n- \rightarrow \text{various hydrocarbons} + \text{char} \quad (4.44)$$

Catalytic pyrolysis using **zeolite catalysts** or **metal catalysts** can selectively produce desired products. **FCC catalysts**(fluid catalytic cracking) favor gasoline-range hydrocarbons, while **ZSM-5 zeolites** produce higher yields of aromatics.

Gasification converts polymers to **synthesis gas** ($CO + H_2$) through partial oxidation at high temperatures (800-1000°C):

$$\text{Polymer} + O_2 + H_2O \rightarrow CO + H_2 + CO_2 \quad (4.45)$$

The synthesis gas can be converted to methanol, diesel fuel, or used as feedstock for new polymer production through **Fischer-Tropsch synthesis**.

Solvolysis uses solvents to break polymer chains through nucleophilic attack. **Alcoholysis**, **hydrolysis**, and **aminolysis**represent different solvolytic approaches depending on the nucleophile employed. These processes typically operate under milder conditions than pyrolysis but require solvent recovery and purification.

Enzymatic depolymerization employs specific enzymes to cleave polymer bonds with high selectivity. **PET hydrolases**discovered in bacteria such as *Ideonellasakaiensis* can completely degrade PET to terephthalic acid and ethylene glycol under mild conditions. **Protein engineering** efforts aim to improve enzyme activity and stability for industrial applications.

Advanced recycling technologies combine multiple approaches for maximum efficiency. **Hybrid processes** may include mechanical preprocessing, chemical depolymerization, and product purification to handle mixed plastic waste streams effectively.

Table 7: Chemical Recycling Technologies for Major Plastic Types

Plastic Type	Primary Technology	Products	Temperature (°C)	Advantages
PET	Glycolysis	BHET, monomers	180-240	High selectivity
PE/PP	Pyrolysis	Mixed hydrocarbons	400-600	Handles mixed waste
PS	Depolymerization	Styrene monomer	300-400	Pure monomer recovery
PVC	Hydrolysis	HCl, organics	250-350	Chlorine recovery
Polyurethane	Glycolysis	Polyols, diamines	180-220	Component recovery

Key Takeaways

This chapter has explored the fundamental principles of thermodynamics, kinetics, and green chemistry, demonstrating their crucial roles in developing sustainable chemical processes and addressing global environmental challenges. The integration of these classical chemical sciences with environmental consciousness provides the foundation for designing tomorrow's chemical technologies.

- **Thermodynamic principles** guide the design of energy-efficient processes by identifying favorable reaction conditions and predicting equilibrium behavior. Understanding free energy relationships enables chemists to design reactions that proceed spontaneously while minimizing energy input and waste generation. The application of activity coefficients and phase equilibria to real systems provides the tools necessary for optimizing separation processes and predicting environmental fate of chemical substances.

- **Kinetic analysis** reveals the pathways and rates of chemical transformations, enabling optimization of reaction conditions for maximum efficiency and selectivity. Transition state theory provides molecular-level insights that guide catalyst design and process improvement. Understanding complex kinetic phenomena such as branched chain reactions and oscillatory behavior helps ensure safe operation of chemical processes and prediction of environmental dynamics.

- **Electrochemical systems** represent key technologies for sustainable energy storage and conversion, offering pathways to integrate renewable energy sources while minimizing environmental impact. Fuel cells and advanced batteries provide clean alternatives to combustion-based energy systems, while electrochemical processes enable efficient chemical transformations under mild conditions.

- **Green polymer technologies** address the critical challenge of plastic waste through design of biodegradable materials and development of chemical recycling processes. Life cycle assessment provides the framework for evaluating environmental impacts and identifying opportunities for improvement throughout the polymer value chain.

The convergence of these fields demonstrates that sustainable chemistry requires understanding of fundamental principles combined with creative application to environmental challenges. Future developments in chemical technology will increasingly depend on this integration of theoretical knowledge with environmental consciousness, driving innovations that benefit both human society and the natural environment.

As chemical processes become more sophisticated and environmental regulations more stringent, the principles explored in this chapter will become increasingly important for developing the next generation of sustainable technologies. The challenge for future chemists lies in applying these principles creatively to design processes that are simultaneously economically viable, environmentally benign, and socially beneficial.

REFERENCES

1. Atkins, P., & de Paula, J. (2018). *Physical Chemistry: Thermodynamics, Structure, and Change* (11th ed.). Oxford University Press.
2. Anastas, P. T., & Warner, J. C. (2000). *Green Chemistry: Theory and Practice*. Oxford University Press.
3. Skoog, D. A., Holler, F. J., & Crouch, S. R. (2017). *Principles of Instrumental Analysis* (7th ed.). Cengage Learning.
4. Nielsen, S. S. (2017). *Food Analysis* (5th ed.). Springer International Publishing.
5. Manahan, S. E. (2016). *Environmental Chemistry* (10th ed.). CRC Press.
6. McMurry, J., & Fay, R. C. (2019). *General Chemistry: Atoms First* (2nd ed.). Pearson Education.
7. Laidler, K. J., Meiser, J. H., & Sanctuary, B. C. (2003). *Physical Chemistry* (4th ed.). Houghton Mifflin Company.
8. Spiro, T. G., &Stigliani, W. M. (2020). *Chemistry of the Environment* (3rd ed.). University Science Books.
9. Voet, D., Voet, J. G., & Pratt, C. W. (2016). *Fundamentals of Biochemistry: Life at the Molecular Level* (5th ed.). John Wiley & Sons.
10. Clayden, J., Greeves, N., & Warren, S. (2012). *Organic Chemistry* (2nd ed.). Oxford University Press.
11. Harris, D. C. (2015). *Quantitative Chemical Analysis* (9th ed.). W. H. Freeman and Company.
12. Bard, A. J., & Faulkner, L. R. (2001). *Electrochemical Methods: Fundamentals and Applications* (2nd ed.). John Wiley & Sons.
13. DeMan, J. M., Finley, J. W., Hurst, W. J., & Lee, C. Y. (2018). *Principles of Food Chemistry* (4th ed.). Springer International Publishing.
14. Schwarzenbach, R. P., Gschwend, P. M., & Imboden, D. M. (2016). *Environmental Organic Chemistry* (3rd ed.). John Wiley & Sons.
15. Lemire, R. J., Berner, U., Musikas, C., Palmer, D. A., Taylor, P., &Tochiyama, O. (2013). *Chemical Thermodynamics of Iron* (Vol. 13a). OECD Nuclear Energy Agency.

16. Zoski, C. G. (Ed.). (2007). *Handbook of Electrochemistry*. Elsevier Science.
17. Cussler, E. L. (2009). *Diffusion: Mass Transfer in Fluid Systems* (3rd ed.). Cambridge University Press.
18. O'Mara, W. C., Herring, R. B., & Hunt, L. P. (2018). *Handbook of Semiconductor Silicon Technology*. William Andrew Publishing.
19. Stevens, M. P. (1999). *Polymer Chemistry: An Introduction* (3rd ed.). Oxford University Press.
20. Wankat, P. C. (2016). *Separation Process Engineering: Includes Mass Transfer Analysis* (4th ed.). Prentice Hall.
21. Holum, J. R. (1990). *Fundamentals of General, Organic, and Biological Chemistry* (4th ed.). John Wiley & Sons.
22. Rodgers, G. E. (2006). *Descriptive Inorganic, Coordination, and Solid State Chemistry* (3rd ed.). Brooks Cole.
23. Connors, K. A. (1990). *Chemical Kinetics: The Study of Reaction Rates in Solution*. Wiley-VCH.
24. Castellan, G. W. (1983). *Physical Chemistry* (3rd ed.). Addison-Wesley Publishing Company.
25. Chang, R., & Goldsby, K. A. (2016). *Chemistry* (12th ed.). McGraw-Hill Education.
26. Mortimer, R. G. (2008). *Physical Chemistry* (3rd ed.). Academic Press.
27. Levine, I. N. (2008). *Physical Chemistry* (6th ed.). McGraw-Hill Education.
28. Moore, W. J. (1972). *Physical Chemistry* (4th ed.). Prentice-Hall.
29. Engel, T., & Reid, P. (2012). *Physical Chemistry* (3rd ed.). Pearson Education.
30. McQuarrie, D. A., & Simon, J. D. (1997). *Physical Chemistry: A Molecular Approach*. University Science Books.
31. Housecroft, C. E., & Constable, E. C. (2010). *Chemistry* (4th ed.). Pearson Education.
32. Shriver, D., & Atkins, P. (2010). *Inorganic Chemistry* (5th ed.). Oxford University Press.
33. Cotton, F. A., Wilkinson, G., Murillo, C. A., & Bochmann, M. (1999). *Advanced Inorganic Chemistry* (6th ed.). John Wiley & Sons.
34. Miessler, G. L., Fischer, P. J., & Tarr, D. A. (2013). *Inorganic Chemistry* (5th ed.). Pearson Education.
35. Huheey, J. E., Keiter, E. A., & Keiter, R. L. (1993). *Inorganic Chemistry: Principles of Structure and Reactivity*(4th ed.). HarperCollins College Publishers.

36. Burgess, J. (1988). *Ions in Solution: Basic Principles of Chemical Interactions*. Ellis Horwood.
37. Richter, P., & Seppelt, K. (2000). *The Chemistry of Hypervalent Compounds*. Wiley-VCH.
38. Cotton, F. A., & Wilkinson, G. (1988). *Advanced Inorganic Chemistry* (5th ed.). John Wiley & Sons.
39. Wells, A. F. (1984). *Structural Inorganic Chemistry* (5th ed.). Oxford University Press.
40. Greenwood, N. N., & Earnshaw, A. (1997). *Chemistry of the Elements* (2nd ed.). Butterworth-Heinemann.
41. March, J. (1992). *Advanced Organic Chemistry: Reactions, Mechanisms, and Structure* (4th ed.). John Wiley & Sons.
42. Carey, F. A., & Sundberg, R. J. (2007). *Advanced Organic Chemistry* (5th ed.). Springer Science+Business Media.
43. Smith, M. B. (2019). *March's Advanced Organic Chemistry: Reactions, Mechanisms, and Structure* (8th ed.). John Wiley & Sons.
44. Streitwieser, A., Heathcock, C. H., &Kosower, E. M. (1992). *Introduction to Organic Chemistry* (4th ed.). Macmillan Publishing Company.
45. Vollhardt, K. P. C., & Schore, N. E. (2014). *Organic Chemistry: Structure and Function* (7th ed.). W. H. Freeman and Company.
46. Solomons, T. W. G., Fryhle, C. B., & Snyder, S. A. (2016). *Organic Chemistry* (12th ed.). John Wiley & Sons.
47. Klein, D. R. (2017). *Organic Chemistry* (3rd ed.). John Wiley & Sons.
48. Bruice, P. Y. (2016). *Organic Chemistry* (8th ed.). Pearson Education.
49. McMurry, J. (2015). *Organic Chemistry* (9th ed.). Cengage Learning.
50. Wade, L. G. (2016). *Organic Chemistry* (8th ed.). Pearson Education.
51. Berg, J. M., Tymoczko, J. L., &Stryer, L. (2015). *Biochemistry* (8th ed.). W. H. Freeman and Company.
52. Nelson, D. L., & Cox, M. M. (2017). *Lehninger Principles of Biochemistry* (7th ed.). W. H. Freeman and Company.
53. Garrett, R. H., & Grisham, C. M. (2016). *Biochemistry* (6th ed.). Cengage Learning.
54. Matthews, C. K., van Holde, K. E., & Ahern, K. G. (1999). *Biochemistry* (3rd ed.). Benjamin Cummings.
55. Zubay, G. L. (1998). *Biochemistry* (4th ed.). WCB/McGraw-Hill.

56. Champe, P. C., Harvey, R. A., & Ferrier, D. R. (2005). *Lippincott's Illustrated Reviews: Biochemistry* (3rd ed.). Lippincott Williams & Wilkins.
57. Devlin, T. M. (2010). *Textbook of Biochemistry with Clinical Correlations* (7th ed.). John Wiley & Sons.
58. Murray, R. K., Bender, D. A., Botham, K. M., Kennelly, P. J., Rodwell, V. W., & Weil, P. A. (2015). *Harper's Illustrated Biochemistry* (30th ed.). McGraw-Hill Education.
59. Rodwell, V. W., Bender, D. A., Botham, K. M., Kennelly, P. J., & Weil, P. A. (2018). *Harper's Illustrated Biochemistry* (31st ed.). McGraw-Hill Education.
60. Pratt, C. W., & Cornely, K. (2013). *Essential Biochemistry* (3rd ed.). John Wiley & Sons.
61. Koolman, J., &Röhm, K. H. (2005). *Color Atlas of Biochemistry* (2nd ed.). Thieme Medical Publishers.
62. Ferrier, D. R. (2017). *Lippincott Illustrated Reviews: Biochemistry* (7th ed.). Wolters Kluwer.
63. Rawn, J. D. (1989). *Biochemistry*. Neil Patterson Publishers.
64. Mathews, C. K., & van Holde, K. E. (1990). *Biochemistry*. Benjamin/Cummings Publishing Company.
65. Stryer, L. (1995). *Biochemistry* (4th ed.). W. H. Freeman and Company.
66. Bailar, J. C., Emeléus, H. J., Nyholm, R., & Trotman-Dickenson, A. F. (Eds.). (1973). *Comprehensive Inorganic Chemistry* (5 vols.). Pergamon Press.
67. King, R. B. (Ed.). (1994). *Encyclopedia of Inorganic Chemistry* (8 vols.). John Wiley & Sons.
68. Pope, M. T. (1983). *Heteropoly and IsopolyOxometalates*. Springer-Verlag.
69. Holm, R. H., Kennepohl, P., & Solomon, E. I. (1996). Structural and functional aspects of metal sites in biology. *Chemical Reviews*, 96(7), 2239-2314.
70. Lippard, S. J., & Berg, J. M. (1994). *Principles of Bioinorganic Chemistry*. University Science Books.
71. Kaim, W., &Schwederski, B. (1994). *Bioinorganic Chemistry: Inorganic Elements in the Chemistry of Life*. John Wiley & Sons.
72. Bertini, I., Gray, H. B., Lippard, S. J., & Valentine, J. S. (1994). *Bioinorganic Chemistry*. University Science Books.
73. Roat-Malone, R. M. (2002). *Bioinorganic Chemistry: A Short Course*. John Wiley & Sons.
74. Ochiai, E. I. (1987). *General Principles of Biochemistry of the Elements*. Plenum Press.

75. Frausto da Silva, J. J. R., & Williams, R. J. P. (2001). *The Biological Chemistry of the Elements: The Inorganic Chemistry of Life* (2nd ed.). Oxford University Press.
76. Cowan, J. A. (1997). *Inorganic Biochemistry: An Introduction* (2nd ed.). Wiley-VCH.
77. Crichton, R. (2008). *Biological Inorganic Chemistry: An Introduction*. Elsevier Science.
78. Thornton, P., & Puddephatt, R. J. (1985). *The Chemistry of the Metallic Elements*. Ellis Horwood.
79. Douglas, B., McDaniel, D., & Alexander, J. (1994). *Concepts and Models of Inorganic Chemistry* (3rd ed.). John Wiley & Sons.
80. Porterfield, W. W. (1993). *Inorganic Chemistry: A Unified Approach* (2nd ed.). Academic Press.
81. Rayner-Canham, G. (2000). *Descriptive Inorganic Chemistry* (2nd ed.). W. H. Freeman and Company.
82. Jolly, W. L. (1991). *Modern Inorganic Chemistry* (2nd ed.). McGraw-Hill Education.
83. Basolo, F., & Johnson, R. C. (1986). *Coordination Chemistry*. Science Reviews.
84. Purcell, K. F., & Kotz, J. C. (1977). *Inorganic Chemistry*. W. B. Saunders Company.
85. Cotton, F. A., & Wilkinson, G. (1972). *Advanced Inorganic Chemistry* (3rd ed.). Interscience Publishers.
86. Sanderson, R. T. (1967). *Inorganic Chemistry*. Reinhold Publishing Corporation.
87. Moeller, T. (1982). *Inorganic Chemistry: A Modern Introduction*. John Wiley & Sons.
88. Mackay, K. M., Mackay, R. A., & Henderson, W. (2002). *Introduction to Modern Inorganic Chemistry* (6th ed.). Nelson Thornes.
89. Wulfsberg, G. (2000). *Inorganic Chemistry*. University Science Books.
90. Sharpe, A. G. (1992). *Inorganic Chemistry* (3rd ed.). Longman Scientific & Technical.
91. Lee, J. D. (1999). *Concise Inorganic Chemistry* (5th ed.). Blackwell Science.
92. Atwood, J. D. (1997). *Inorganic and Organometallic Reaction Mechanisms* (2nd ed.). Wiley-VCH.
93. Jordan, R. B. (1998). *Reaction Mechanisms of Inorganic and Organometallic Systems* (2nd ed.). Oxford University Press.

94. Tobe, M. L., & Burgess, J. (1999). *Inorganic Reaction Mechanisms*. Addison Wesley Longman.
95. Wilkins, R. G. (1991). *Kinetics and Mechanism of Reactions of Transition Metal Complexes* (2nd ed.). VCH Publishers.
96. Espenson, J. H. (1995). *Chemical Kinetics and Reaction Mechanisms* (2nd ed.). McGraw-Hill Education.
97. Basolo, F., & Pearson, R. G. (1967). *Mechanisms of Inorganic Reactions: A Study of Metal Complexes in Solution*(2nd ed.). John Wiley & Sons.
98. Langford, C. H., & Gray, H. B. (1965). *Ligand Substitution Processes*. W. A. Benjamin.
99. Eigen, M., & Wilkins, R. G. (1965). The kinetics and mechanism of formation of metal complexes. *Advances in Chemistry Series*, 49, 55-80.
100. Margerum, D. W., Cayley, G. R., Weatherburn, D. C., & Pagenkopf, G. K. (1978). Kinetics and mechanisms of complex formation and ligand exchange. *Coordination Chemistry*, 2, 1-220.